Stars and Clusters

Stars and Clusters

CECILIA PAYNE-GAPOSCHKIN

HARVARD UNIVERSITY PRESS
Cambridge, Massachusetts
London, England · 1979

Library of Congress Cataloging in Publication Data

Gaposchkin, Cecilia Helena Payne, 1900–
    Stars and clusters.

    (The Harvard books on astronomy)
    Includes index.
    1. Stars. 2. Stars—Clusters. I. Title.
II. Series.
QB801.G24   523.8   79-4472
ISBN 0-674-83440-2

For my husband,
that "bright particular star"

Contents

Stars and Clusters

All the World's a Stage—The Galaxy

The cosmic drama is played out on a vast scale in time and space. The mind, moving "on wings as swift as meditation," perceives today the events that were passing at the bounds of the observable universe several thousand million years ago—events of which the news has just reached us, carried with the speed of light. Some of these events seem foreign to us, like the customs of a distant land. They may spring from a different environment, a different history. They may stem from a different stage of development. We catch glimpses of the drama that is being played out on the fringes of the world, and they carry a suggestion of the Theater of the Absurd.

The piece that I intend to present has a more modest scope. It will be played on familiar ground, the tiny stage of our own galaxy. Ours is one of many hundreds of millions of stellar systems, a tiny sample of the universe, and is certainly not in any way unique. Some galaxies are larger, brighter, more populous. Even more are smaller and fainter and have fewer members. A flattened, rotating complex of stars and interstellar material, our galaxy contains perhaps a million million stars and extends about 30 kiloparsecs from edge to edge. The universe that we have studied stretches about fifty thousand times as far and contains perhaps ten thousand million galaxies. Our own system is a mere drop in the ocean.*

CHAPTER 1

* In expressing distances, velocities, and masses it is convenient to make use of units that do not involve unwieldy numbers. The basic unit of distance in stellar astronomy is the *parsec*, roughly 3×10^{13} kilometers. One parsec is the distance of a star whose apparent position differs by two seconds of arc when viewed from opposite sides of the orbit of the earth around the sun. In terms of the popularly used but verbally deceptive *light year* (which is of course a distance, not a time interval), one parsec equals 3.263 light years. The enormous

Figure 1.1.   Seven typical galaxies. Top row, left to right: NGC 628 (Messier 74), viewed almost face-on; NGC 4725, viewed at an angle; NGC 4565, viewed on edge. All three are spiral galaxies. Bottom row, left to right: the giant elliptical galaxy Messier 87, surrounded by an aura of globular clusters; the small elliptical galaxy NGC 205, a companion of the Andromeda spiral Messier 31; (above) the irregular galaxy 4449; (below) the peculiar irregular galaxy Messier 82, which shows evidence of a gigantic explosion in the far past. (Photographs by Hale Observatories.)

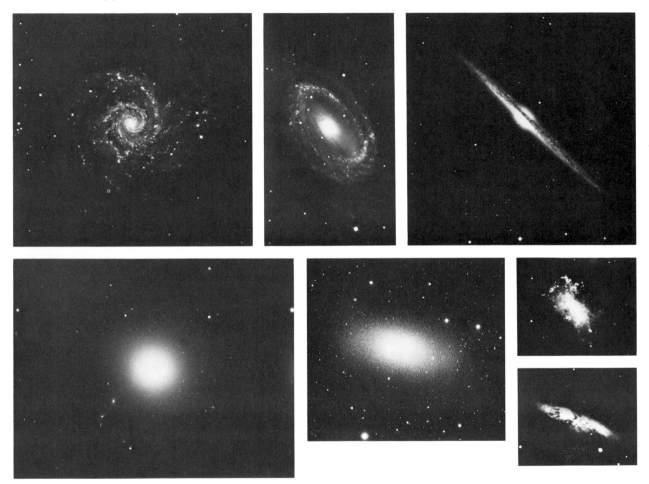

scale of interstellar distances makes even the parsec an unwieldy unit, and the *kiloparsec* (1000 parsecs, or about 3×10^{16} kilometers) is more convenient. For intergalactic distances we may employ the *megaparsec* (a million parsecs).

Velocities are expressed in kilometers per second.

Masses of stars are conveniently given in terms of the sun's mass, 1.99×10^{33} grams. On this scale, all known stellar masses lie between 100 and 1/100.

That our galaxy is not in any way unique does not imply that it is a typical sample of the universe. Details of its structure and composition can do little more than furnish clues to the workings of one stellar system and to the way in which it and its components have developed. That a number of other stellar systems have similar components and seem to be traveling the same road gives a certain generality to our inquiry; more than this we cannot claim.

Galaxies come in a variety of shapes as well as sizes. Figure 1.1 shows seven that illustrate the recognized types: spiral, elliptical, and irregular. There is an understandable tendency to choose, for illustration, systems that present a neat symmetrical pattern, and I have not resisted it. But all galaxies have their idiosyncracies. Many spirals are lopsided or distorted, especially (though not always) when other galaxies are close by. It is as great a mistake to think of a galaxy as a neatly convoluted spiral as it is to picture a star as a glowing billiard ball. Both are developing, both are in the grip of inescapable physical conditions. Neither galaxy nor star was the same in the past as it is now, nor will it be the same in the future. Both "have their exits and their entrances." The task I have set myself is to trace the entrances and exits of stars on the little arena of our own galaxy and to describe the behavior of each as he "struts and frets his hour upon the stage."

The majority of galaxies are elliptical, ranging from the enormous Messier 87 (fig. 1.1) to small, faint ones with few members, such as the Sculptor system.* Irregular galaxies seem to be the rarest. Here, too, we find bright and populous ones like the Large Magellanic Cloud and tiny ones like Leo II. Our own galaxy is a spiral of intermediate size. There are enormous spiral systems like Messier 31 in Andromeda, larger than our

---

* A few galaxies and clusters have familiar names, such as the Great Nebula in Andromeda (actually not a nebula but a galaxy) or the Hyades. Most are known by serial numbers that have been assigned to them in specialized catalogues; these numbers are preceded by letters such as M, NGC, or IC that identify the catalogue.

M stands for Charles Messier, the French astronomer who, over a century and a half ago, compiled a catalogue of non-comets, that is, objects to be disregarded by seekers of those tiny interlopers of the solar system. In the process Messier produced one of the first lists of galaxies and star clusters.

NGC stands for the *New General Catalogue* compiled by John Louis Emil Dreyer, which, like Messier's list, is a motley collection of galaxies, clusters, and true nebulae. It is no longer new, having been compiled in 1888 from several earlier catalogues, but the numbers assigned in it are still the recognized names of the objects. IC stands for the two additional *Index Catalogues*, which amplified the *New General Catalogue* in 1895 and 1908.

A large number of galaxies and clusters have been discovered and studied since the *New General Catalogue* and the *Index Catalogues* were compiled, and they are variously named. Leo II is the second faint galaxy in the constellation Leo. Clusters such as Hogg 5 and Lynga 6 are identified by the number given to them by the compiler of the relevant catalogue.

own. Smaller spirals there are too, like Messier 33 in Triangulum and Messier 51, the Whirlpool, but there are few very small ones. The spiral galaxies look as though they are spinning, as indeed they are.

Galaxies are not distributed uniformly in space; they tend to cluster. Some clusters contain thousands of members, some hundreds. Our own is one of a small company, the so-called Local Group, which contains spirals, ellipticals, and irregulars—about twenty members in all. Even within the Local Group there is a tendency to form smaller, more compact groups. Our own galaxy has two "puppies," the Large and Small Magellanic Clouds. The Andromeda spiral, itself a member of the Local Group, is associated with at least four small elliptical systems. This tendency to form groups within groups will appear repeatedly as we explore.

Ironically, we cannot produce a detailed portrait of our own galaxy. Not only are we inside it but we are in the central plane, which (as in all such spiral systems), is thick with interstellar material—dust, molecules,

Figure 1.2.   Artist's impression of the galactic system, as seen from the north galactic pole. It was prepared from radio observations of the emission from neutral hydrogen atoms; the density contrasts are greatly exaggerated in order to display the general shape of the spiral structure. The observations were made by the Netherlands Foundation for Radio Astronomy in Leiden and by the Division of Radiophysics, Commonwealth Scientific and Industrial Research Organization, Sydney, Australia. (Diagram by G. Westerhout, reproduced with his kind permission.)

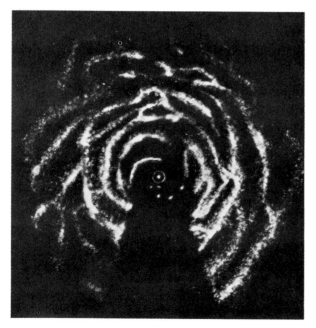

Figure 1.3.   The two nearby galaxies Messier 31 (the Andromeda spiral) and Messier 81 (the Ursa Major spiral). In form and population they resemble our own system. Both are seen at an angle; if viewed face-on they would present roughly circular outlines. (Photographs by Hale Observatories.)

atoms, and ions. More than half the system is hidden from optical study by absorbing material, and much of the rest, if not concealed, is obscured. Observations with the radio telescope tell us more, for radio waves can pierce the obscuration and make it possible to map a large fraction of the Galaxy. Radio observations, surveying the Galaxy by the light of atomic and molecular clouds, have amplified the optical picture to show the swirling pattern that is loosely described as "spiral structure." Figure 1.2 shows the distribution of neutral hydrogen gas in the central layer of our system. This distribution has been deduced from the observed velocities of the gaseous clouds by means of a sophisticated model of the distribution of mass within the Galaxy and the effect of this distribution on the rotation of the system. The results obtained from radio astronomy amply justify the description of our galaxy as a spiral, even though the pattern is fragmentary and disjointed. Few of the recognized spiral galaxies, in fact, show neatly convoluted spiral "arms." Figure 1.3

shows two well-known systems that are probably very like our own: the Andromeda spiral and the Ursa Major spiral.

From radio observations and optical studies of the distribution and motion of stars we arrive at a general picture of our galaxy. The central layer is a greatly flattened rotating system, within which lies a swirling pattern of glowing gas streaked with absorbing material. The outer swirls trail behind the general direction of rotation: their angular velocity is smaller than that of the "spiral" features nearer the center. Similar spiral patterns —though not for the same reason—are displayed by water swirling out of a sink. The brightest, hottest stars share the rotational motion and tend to lie in the swirls of gas within the flattened layer. These are among the youngest stars. Older stars are not so closely confined to the central plane nor to the swirling arms. Many of them make up a thickened disk that grows less populous toward its outer edges and shares the rotational motion. The very oldest stars are distributed in a more spherical system that is concentrated to the galactic center and forms a sort of halo around the whole. These stars depart more and more from the rotation of the central layer and can thus be recognized by their velocities as well as by their positions; they show no sign of spiral arrangement. So we recognize a thin rotating central layer and swirling arms, a thickened disk into which the central layer merges, and a halo whose members are distinguished by both distribution and motion.

The complex nature of the Galaxy—spherical halo, disk and central flattened layer—is emphasized by differences of gross properties. Hot stars of high luminosity are confined to the central layer. The halo contains many cooler stars of lower luminosity. The result, more easily seen in distant spiral galaxies that can be surveyed as a whole, is a difference of color (an index of the temperatures of the component stars) between halo and central layer. Equally noteworthy is a gradient in chemical composition between the center and edge of the flattened system: metallic atoms grow progressively rarer with distance from the center. Even more striking is a paucity of metallic atoms in the halo as compared with the disk and the central layer.

The constituents of the Galaxy display many levels of complexity. Simplest are the ions and the atoms, such as the hydrogen of the swirling clouds. Then come the molecules, some quite simple like the molecules of hydrogen (H_2), carbon (C_2), carbon monoxide (CO), the cyanogen radical (CN), the hydroxyl radical (OH), and some fairly complex like methane (CH_4), methyl alcohol ($CH_3 \cdot OH$), ethyl alcohol ($CH_3 \cdot CH_2 \cdot OH$), formaldehyde ($CH_2 \cdot O$), acetaldehyde ($CH_3 \cdot CHO$), and formic acid ($H \cdot COOH$), all of which have been revealed by the radio telescope. The most complex molecule yet detected in interstellar space is cyano-octate-trane (HC_9N). More than thirty interstellar molecules have been observed. But neither atoms nor molecules can be responsible for the massive absorption that streaks and pervades the central plane. Matter must also be there in the form of dust and larger solid particles, very likely silicates and graphite. The light of distant stars is polarized, and solid particles must be responsible for this. Virtually all the interstellar material lies near the central plane of the Galaxy and probably shares the general rotation.

Next in the scale of complexity come the stars. Not counting the sun, those visible to the unaided eye range from the brilliant Sirius to objects about a thousand times fainter. Our system of expressing their brightness goes back to Claudius Ptolemy, the second century Alexandrian astronomer, who assigned the numbers 1 to 6, first magnitude stars being the brightest. In consequence of the so-called "law" of Gustav Theodor Fechner, "In order that the intensity of a sensation may increase in arithmetical progression, the stimulus must increase in geometrical progression," Ptolemy's scale is logarithmic. Equal steps in magnitude correspond to equal ratios in brightness: a difference of five magnitudes corresponds to a ratio of a hundred in brightness. The modern telescope permits the observation of stars of at least the twenty-third magnitude, more than a thousand million times fainter than Sirius. The directly observed brightness is known as the *apparent magnitude*. When the system was placed on an accurate basis, zero and negative magnitudes were found to be necessary for a few stars such as Sirius and Canopus. The apparent magnitude of the sun is -26.72.

Even to a casual glance the stars are not all of the same color. Sirius is brilliantly white, Antares ("rival of Mars") is reddish. Two stars that seem of the same brightness to the eye, which is most sensitive to yellow light, will not make the same impression on an unsensitized photographic plate, which responds to blue light: the blue star will appear the brighter. The difference can be accentuated by observing through filters that transmit light with small ranges of color. When the brightness of a star is to be accurately expressed, the color in which it has been observed must be specified. Standard filters define the V (visual) and B (blue, which corresponds roughly to the photographic response) systems. Other filters isolate the U (ultraviolet) and a number of infrared systems. The color of a star is then expressed by the difference between the magnitudes that have been measured in the different systems, which are so adjusted that all of them coincide for a star of about the color of Sirius. The resulting number is known as the *color index*—for example $B - V$, where B is the magnitude measured on the *B* system, *V*, that on the V system. For Achernar (α Eridani) V = 0.47, *B* = 0.28, and the color index is −0.19; for Aldebaran (α Tauri), V =0.86, *B* = 2.39, and the color index is +1.53. Color index is a measure of surface temperature: the redder a star, the cooler it is.

Because stars are not all at the same distance, their apparent magnitudes do not represent their true brightnesses, or *luminosities*. If the distance of a star is known, its *absolute magnitude* (by convention, the magnitude it would have if it were at a distance of ten parsecs) can be found. The range of absolute magnitudes of stars (other than those undergoing explosions) is from about −8 to at least +16.

Figure 1.4, which represents the luminosities, color indices, and temperatures of a group of stars chosen because they are bright and familiar, emphasizes that no simple relation exists between these quantities. It is indeed an understatement of the enormous range of their properties. For convenience in expressing this variety they are referred to several groups

Figure 1.4. Brightness and surface temperature for representative stars in the Galaxy. Only the more luminous stars are included; the known faint stars extend to near absolute visual magnitude +20, with temperatures down to about 3000°.

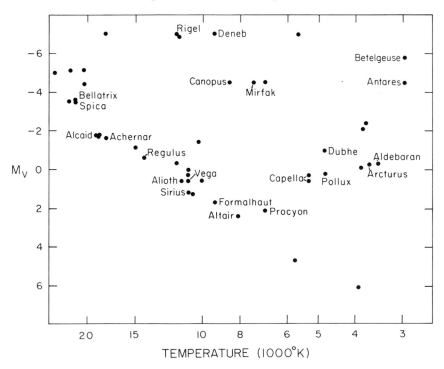

Figure 1.5. Radius and color for stars of various luminosity classes. Radii are expressed as multiples of the solar radius. The broken line for class V represents a correction to the magnitudes (and sizes) to allow for absorption by metallic oxides.

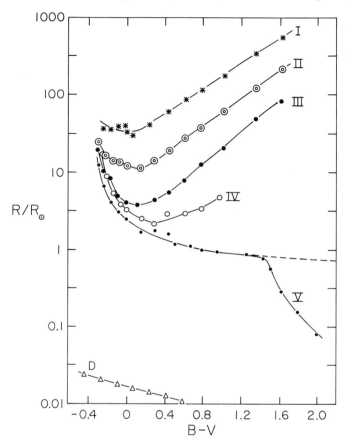

(luminosity classes) designated by Roman numerals: supergiants (I), bright giants (II), giants (III), subgiants (IV), main-sequence or dwarf stars (V), and white dwarfs (D). It must be emphasized that, except for D, the groups are not sharply divided but grade into one another. At any one temperature there is a continuous change of properties as we pass from I to V. Stars of luminosity class I are the largest and rarest, those of luminosity class V the smallest and most frequent. Most stars, in fact, are of class V, and the relationship that it defines is known as the *main sequence*.

Figure 1.5 illustrates the relationship between the average sizes of stars of various color indices ($B - V$) and luminosity classes. Diameters range from a thousand times to one hundredth that of the sun. They have been calculated from the luminosities and color indices, which have been converted into surface temperatures by an application of the law of radiation formulated by Max Planck. If the stars were perfect radiators ("black bodies"), the energy radiated by unit area of surface (such as a square centimeter) would be proportional to the fourth power of the temperature, according to the "Stefan–Boltzmann law" derived by Josef Stefan and Ludwig Boltzmann. Therefore a cool star must have a larger surface area than a hot one if its total radiation is to be the same. Actually the stars are not black bodies, but for most of them the formula gives a good approximation to their sizes. The most conspicuous deviation is that for the cooler stars, and results from the absorption of their light by chemical compounds in their atmospheres. The broken line in figure 1.5 for the coolest stars of luminosity class V was obtained by allowing for this effect.

Brightness, temperature, and size are the obvious external characteristics of stars. But their most important property, the one that principally determines their physique, their course of development, and their ultimate fate, is their mass. Stars of small mass develop slowly, and there are limits to what they can achieve. Stars of large mass develop swiftly, and there are limits to the length of their career, which may end in a blaze of glory. In chapter 3 we shall examine the way in which a star's mass influences its life history.

Early studies of the spectra of the stars established that they all consist of the same chemical elements, the same atoms that are familiar to us on the earth. It was, therefore, easy to come to the erroneous conclusion that their chemical compositions are identical. We know today that this is far from being the case. Chemical composition is second only to mass in determining the course and speed of a star's development. It is also capable of furnishing important clues to origin and behavior.

What I have called behavior—the fluctuations of brightness, the variations of temperature and spectrum that characterize the variable stars—is the most obvious and provocative stellar property. Most stars appear to be steady and unwavering, though the sun should teach us a lesson here. Despite the fact that our own star—so close to us, so easy to study in detail—can be seen to seethe with activity, if it were as distant as α Centauri (a binary system whose brighter member is very nearly the sun's twin), none of this activity would have been noticed. And the α Centauri system is our nearest neighbor among the stars.

Unlike the sun, the variable stars change in brightness, spectrum, and color, some rhythmically, others spasmodically. There was a tendency not so long ago to regard them as pathological stars. Today we see them as stars that are reacting naturally to some critical phase of their develop-

Figure 1.6.   Relation of the principal types of variable star to absolute visual magnitude and color index $(B - V)$. The domains of nonvariable stars in open clusters and globular clusters are lightly shaded in the left and right halves of the diagram.

Left: The irregular pre-main-sequence *flare stars* occupy an area above and to the right of the main sequence. The pulsating *β Canis Majoris stars*, with very short periods of the order of hours, lie near the bright tip of the main sequence; lower, near the main sequence, are the pulsating *δ Scuti stars*, also of very short period. Brighter and redder than the latter, the *Cepheid variables* have longer periods of the order of days, and more regular variations.

Right: Brightest and reddest are the *long-period variables or Mira stars*, pulsating with periods of the order of months. The *W Virginis stars* and the *RV Tauri stars* pulsate in periods comparable to those of Cepheid variables. The fainter, bluer *RR Lyrae stars* pulsate in periods of hours. The explosive variables include the *novae* and *dwarf novae*; they undergo violent outbursts, the former at intervals of decades to many thousand years, the latter at intervals of weeks or months.

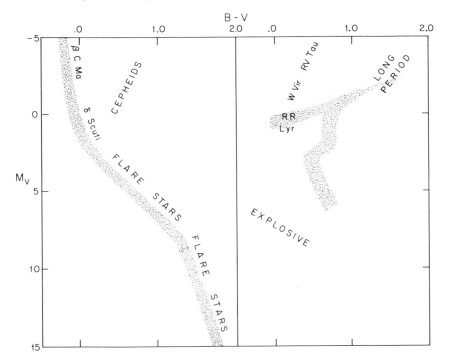

ment, and thus we can place them in the sequence of stages of the stellar lifetime. Indeed, one star "in his time plays many parts." Yesterday a blue supergiant, today a Cepheid variable, tomorrow perhaps a supernova. Yesterday a star like the sun, today a Mira variable, tomorrow perhaps a planetary nebula. The more prominent types of variable stars are shown schematically on a luminosity–color background in figure 1.6. They will appear in turn as we span the stellar lifetime, and their behavior will be described. Here they take only a preliminary bow.

Intimate unions can be a determining factor in the life of a star: today it may be a close binary, tomorrow perhaps a nova. A survey of the stars nearest to us reveals a striking fact: many of them, at least half, are closely associated with other stars. In some cases (*visual binaries*) both stars can be observed moving in orbits around one another. In other cases, though only one star is visible, an apparently sinuous motion across the sky can be interpreted as a combination of linear motion and orbital motion around an unseen companion (*astrometric binaries*). Or the stars may be optically inseparable but may be inferred to be binary from periodic Doppler shifts* of their spectrum lines, evidence of orbital motion (*spectroscopic binaries*). These periodic changes of velocity can reveal a binary even though, because of disparity in brightness, only one of the spectra is observable. Finally, the stars may pass alternately in front of one another (*eclipsing binaries*). Binary stars are the backbone of our knowledge of stellar properties, for they alone can have their masses directly measured.

While a large proportion of stars are double, many are multiple. Castor (α Geminorum) has six components, Albireo (β Cygni) has at least three, and the Trapezium in Orion (θ Orionis) has four components, two of which are eclipsing binaries and the other two probably also double. The list could be multiplied indefinitely, even among the naked-eye stars (see chapter 5).

Binary and multiple stars may form the nuclei of much larger stellar aggregates. In chapter 6 we shall consider NGC 6231, a cluster of bright stars in the constellation Scorpio. Several of its brightest members are binary stars. The cluster itself is quite compact and is one of the nuclei of a great stellar association, a group of young stars that are associated in space.

Associations are not so compact as to form obvious clusters; they are loose groups of stars that have common properties and often contain compact clusters as nuclei. All of them bear the marks of extreme youth. They will form the subject of chapter 4.

More compact than associations are the *open clusters*. They may con-

---

* The change in the pitch of sound as the speed of the source (for example, a passing train) changes from approach to recession is familiar to everyone. The vibrations are crowded together as the source approaches, spread out as it recedes, and the result is a drop in pitch. Light vibrations from a moving source are similarly affected. Those from an approaching source are crowded together, so the light is apparently blued; from a receding source it is apparently reddened: the objects "show their red taillights." The resulting change in the wavelength of spectrum lines coming from the source is measured by a spectroscope, and since the velocity of light is known, the velocity of the source can be deduced.

tain from a few dozen to many hundred stars; and, unlike the associations, they are probably almost all held together by the mutual gravitation of their members. They have a variety of ages, from extremely young to very old indeed.

Very young clusters and associations probably differ only in the circumstances of their formation. A populous, compact cluster is held together by the mutual gravitation of its members, whereas the gravitational bond between the members of a large, loose association is necessarily smaller. Both are subject to disruption by the effects of the star fields through which they pass, but the gravitationally bound clusters resist dispersion better than the loosely held associations. For this reason only the very youngest associations still survive as recognizable groups. Even the clusters must gradually lose members, and we shall see this process in action when we examine the Hyades.

A distant cluster looks like a cluster; a very nearby one cannot be recognized by its appearance. An example is furnished by the Ursa Major group, detected as a cluster only because the motions of its individual stars have been measured and found to be parallel. Actually we are inside it. All the known moving groups are close to us, otherwise they would not have been detected. There is no hard-and-fast distinction between moving groups and clusters. Most of the known moving groups are sparsely populated and are probably not held together so firmly by mutual gravitation as the more compact clusters. But the members of the hierarchy of stellar groups that makes up the Galaxy are closely interwoven. A gravitationally bound cluster often forms the nucleus of a moving group or an association; double or multiple stars, still more closely bound, may form the nuclei of open clusters.

The open clusters include some of the most conspicuous objects. The Hyades, the Pleiades, Praesepe, Coma Berenices, κ Crucis (the Jewel Box), and Messier 11 leap to the eye in the sky or on a photograph. But other apparent clusterings, especially in regions rich in stars, turn out on closer examination not to be true groups.

The youngest open clusters are very close to the central plane of the Galaxy and tend to lie in the swirls that are marked out by the bright hydrogen clouds. The older open clusters do not conform so closely to this pattern, and the very oldest show no such pattern at all, though they too are moving in conformity with the rotational motions of the inhabitants of the central plane. Several hundred open clusters are known. Many more must have remained undiscovered, for open clusters tend to lie in obscured regions and those on the far side of the galactic center are unobservable. One may guess that there are at least a thousand in our system.

The most striking stellar groups in the Galaxy are the *globular clusters*, of which over a hundred are known. Perhaps there are two hundred in all. They are compact, are usually centrally condensed, and can contain more than a million stars, though many are far less populous. Their distribution and motion place them unequivocally as extreme representatives of the halo population. They are concentrated to the center of the Galaxy, where the remaining undiscovered ones must be situated. They are probably among the oldest denizens of our system and will set the scene for our final chapters.

Associations, moving groups, clusters, and double stars furnish touchstones for our ideas about stellar development, for the properties of the members of any one cluster are not distributed at random. Within each cluster the brightness, color, and size of the stars follow a specific pattern, and the patterns displayed by individual clusters can be arranged in a coherent sequence. This observation fortifies the conviction that clusters are not haphazard collections of stars but are groups united by a common origin and similar history.

This conviction has in turn guided the theoretical work that has so successfully traced the course of stellar development (see chapter 3). These researches, stemming from an understanding of the sources of stellar energy, have made it possible to associate the patterns displayed by the clusters with age and stage of development. It is my aim to lay emphasis on observed phenomena. But I cannot avoid leaning heavily on the current theoretical picture of stellar development which has successfully simulated so many of the features of the actual participants in the drama. Theory provides a convincing account of the springs of their behavior, and as such I shall present it.

On the cosmic scale our galaxy is small. But it is very complex, a hierarchy that ranges from ions and atoms through stars to clusters and associations. Many other galaxies are very like it, not only in form but in content. The Andromeda spiral, Messier 31, displays interstellar clouds and a great variety of stars; it, too, has a central layer, a disk, and a halo; it, too, is rotating. It contains great associations, open clusters, and globular clusters, many of which resemble those in our galaxy, and its population of Cepheid variables and novae is very similar to ours. So we may well feel that conclusions drawn from a study of our own system will be valid for Messier 31 and for other similar galaxies.

But we must not go too far. In our very backyard are the Large and Small Magellanic Clouds, which differ from our galaxy and from each other in all sorts of ways: in structure, in the characteristics of their Cepheid variables and star clusters, possibly in their chemical composition as well. Not far away is the irregular galaxy Messier 82 (a companion of Messier 81; see figs. 1.1 and 1.3), once apparently the seat of a gigantic explosion. It is surrounded by such a sea of dust that not a single star is discernible, though the light that filters out through the dust cloud is that of stars, and so stars must be there. The tiny elliptical Sculptor system has a variable star population quite unlike our own. Galaxies such as these have a different tale to tell. They remind us of the limited scope of our inquiry: our sample is not complete.

New techniques are not merely pushing back the limits of the observable universe but are opening up unexplored vistas: distant, compact systems that seem quite unlike those that we know and live in and think we understand. The astronomy of galaxies seems about to enter upon a new phase. At this crucial moment let us turn to take a look at the parochial conditions of our own small, familiar galaxy.

Merely Players—Meet the Hyades

Taurus, the Celestial Bull, is the harbinger of spring. Bestriding the Zodiac,

> candidus auratis aperit cum cornibus annum
> Taurus

("when the gleaming Bull with golden horns ushers in the year," Virgil, *Georgics* 1.217), he furnishes a fitting introduction to the cosmic drama. His image, handed down from antiquity, depicts a half-creature, "nube candentes humeros amictus" ("his shining shoulders shrouded in a cloud," Horace, *Odes* 1.2). A V-shaped group of stars outlines his muzzle, accented by the brilliant Aldebaran (fig. 2.1). He is rich in astronomical interest. Within his borders is a nest of newly-born stars wreathed in interstellar obscuration. The tip of one horn marks the position of the Crab Nebula, site of a supernova whose remains survive as a pulsar. He includes two of the most conspicuous of star clusters, the Pleiades and the Hyades, the latter occupying his muzzle. His horns touch the edges of the Milky Way. As we look toward him we have our backs to the center of the galactic system and our faces to the anticenter.

In this direction the observable Galaxy can be traced to a distance of about 6 kiloparsecs by means of a few distant star clusters. The Hyades lie in the foreground, nearest to us of all clusters, at a mere 40 parsecs (fig. 2.2). They appear to be rather far from the Milky Way: their galactic latitude is $-23.90°$, but this is an effect of perspective, a result of their being very close to us. Actually they are only 16 parsecs south of the galactic plane.*

---

* When the position of something within the Galaxy is to be described, it is convenient to use a system of coordinates similar to the latitude and longitude used to locate places on the

Figure 2.1.   The constellation Taurus, the Bull, from Johann Bayer's *Uranometria*, 1603. (The beautiful engravings in this early star atlas were made by Albrecht Dürer.)

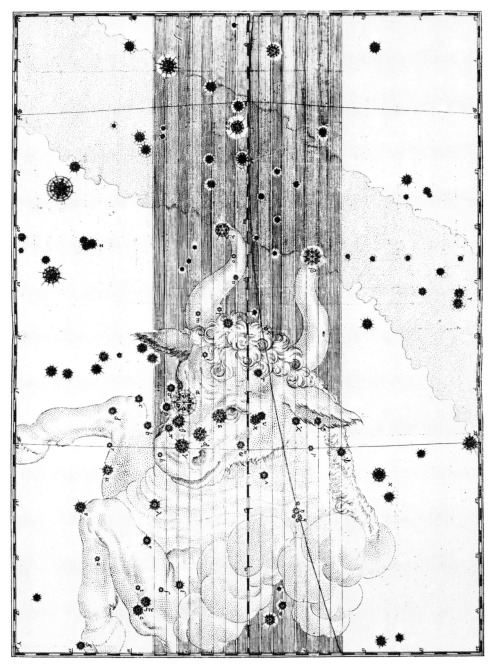

Figure 2.2.   The Hyades cluster. (Photograph by Harvard Observatory.)

---

earth's surface. Terrestrial latitudes are referred to the earth's equator; galactic latitudes are similarly referred to the course of the Milky Way, regarded as the equator of the Galaxy and marked by an imaginary line on the celestial sphere. On the earth, longitudes are referred to an arbitrary line, the prime meridian (fixed by historical considerations). In the Galaxy, they are now counted around the galactic equator from the direction of the center of the galactic system. These galactic coordinates have been determined by means of the radio telescope, the instrument best adapted to circumvent the obstruction of light by the absorbing matter that lies thickest in the central layer.

The central line of the Milky Way, the galactic equator, defines the *galactic plane*, which passes through the center of the Galaxy. The apparent position of an object within the Galaxy is described by its galactic latitude and longitude: in the case of the Hyades, $-23.9°$ and $179.1°$. The cluster is very close in longitude to the galactic *anticenter*, whose longitude is $180°$, the longitude of the center being $0°$. The actual position of an object and its distance from the galactic plane can be found if the distance from us is known. By convention, lati-

When we reflect that our galaxy stretches at least 6 kiloparsecs in the direction of the Hyades, it is clear that all the stars seen within the bounds of the cluster cannot be members of it. Some of them, particularly the faint ones, will be more distant; some will lie in the foreground. The bright Aldebaran, most conspicuous of all, is probably not a member of the cluster. If we wish to make a census of the Hyades, a criterion for membership must be established.

All the members of a permanent group of stars share a common motion. A group like the Hyades has been circulating about the Galaxy for about one hundred million years. By the rule of thumb that one kilometer per second equals approximately one parsec per million years, if a star's motion deviated by one kilometer per second from that of the group, in one hundred million years it would have moved away from the group by 100 parsecs, about seven times the cluster's radius. Thus, it is possible to pick out members of the group by making an accurate survey of their motions.

Because the Hyades are the nearest of all clusters to us, they furnish the cardinal data to which all other clusters in their turn can be related. They are near enough for their motions to reveal the effects of perspective, so that they seem to be converging to a point rather than moving in strictly parallel paths across the sky (fig. 2.3). If, in addition to apparent motion across the sky, radial velocity also can be measured, the true motion in space can be determined.* The perspective effect will then permit a geometrical determination of distance, as well as a determination of the spatial arrangement of the stars.

The exacting and punctilious work that has gone into measuring the motions of the Hyades and sorting out the true members of the cluster can only be appreciated after a first-hand study of the laborious investigations that have extended over many decades. For decades must elapse before the changes in the observed positions have become large enough to be accurately measured. Subtle and elusive corrections must be applied. The picture of the Hyades cluster that has resulted from these investigations shows about four hundred stars confined within a radius of about 14 parsecs and situated about 40 parsecs from the sun. This picture will be refined with the passage of years, for the accumulation of data still con-

---

tudes north of the plane carry a plus, those south of the plane a minus sign; north and south are sometimes described as above and below the plane. Our sun lies very slightly north of the galactic plane.

   * The velocity of an object in our direction, determined from the Doppler shift, is known as its *radial velocity*. By convention, velocity in our direction carries a negative sign, velocity away from us, a positive sign.

Figure 2.3. Diagram showing the motions of the brighter Hyades and the point toward which their parallel motions appear to converge. (Adapted from a diagram by H. G. van Bueren.)

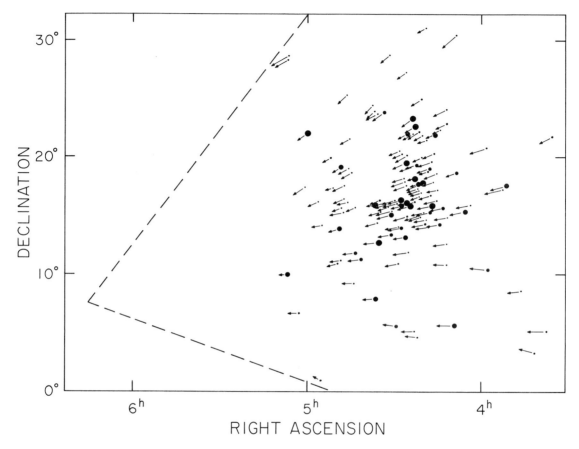

tinues. More powerful instruments and an increasing baseline will add to the number of accepted members and verify or reject doubtful ones.

Today, nevertheless, we have a rather complete census of the Hyades cluster and a good geometrical determination of its distance. One source of uncertainty that besets the derivation of the distance of many clusters is fortunately absent here: the Hyades stars are so close to us that their brightness is not observably diminished by interstellar absorption. Such absorption would not, of course, affect the geometry of determining the distance, but it would falsify the absolute brightness of the member stars by reducing their apparent brightness. We can use the data on the Hyades with confidence to survey the absolute properties of the member stars, whereas for most other clusters it is necessary to make an allowance for interstellar absorption, an allowance that is attended with some difficulty for many of them.

The accepted members of the Hyades cluster (those whose motions conform to the rather strict limits imposed by the investigations just described) are shown in figure 2.4 by large, heavy dots; probable members are indicated by circles. Other stars of known color that lie within the area of the sky covered by the Hyades are shown by small, light dots. A glance at the distribution of these other stars emphasizes that all stars within the apparent boundaries of the cluster are not members. Those that do not conform to the common motion do not share the special pattern of luminosity and color that is characteristic of those that do. It is profitable to reflect on this picture when we examine similar diagrams for clusters too distant to be sorted out by means of measured motions. There will always be some interlopers, though of course there will be fewer for clusters of smaller apparent size.

The stars of the Hyades display a wide range in brightness and also in color. However, figure 2.4 shows a remarkable neatness: stars of a given color are confined to a small range in luminosity. Most stars occupy a strip that runs (top to bottom) from bright to faint and (left to right) from bluer to redder. Four of the bright members, almost the same in luminosity and color, stand out from this pattern, and so do several faint ones. Thus, at first glance the Hyades can be separated into *main-sequence stars* (those making up the diagonal strip), giant stars (the four isolated red ones), and white dwarfs (the isolated faint blue ones).

A similar pattern is shown by every known cluster. Many such patterns (known as *color–magnitude diagrams*) are less well defined than that of the Hyades. This may result from uncertainties in applying the correction for interstellar absorption, which may not be uniform over the face of the cluster (or within it), and affects both color and apparent brightness. Or it may reflect uncertainties in deciding which stars are members of the cluster: when they are near enough, their motions may provide a criterion, and when they are bright enough, their radial velocities will do so. But for faint and distant clusters we lack both of these data, and membership must be assigned on photometric or statistical grounds. All these uncertainties tend to blur the observed pattern. Nevertheless, even when due allowance is made for them, the fact remains that many clusters have far less well-defined color–magnitude diagrams than the Hyades, especially the very young ones.

The color–magnitude patterns shown by different clusters are far from identical, but they do have a common ground in their main sequences. Most stars in the Galaxy lie on the main sequence, the diagonal strip in the color–magnitude diagram that runs from bright blue stars to faint red

Figure 2.4.   Apparent visual magnitude (m_v) and color ($b − v$) for stars in the area of the Hyades. Large dots show undoubted members of the cluster; circles show probable members. Small dots show stars whose motions indicate that they are not members of the cluster. Note the four red giants (in a row above the diagonal strip) and the four white dwarfs (large dots and circle scattered below the diagonal). There are many ways to describe the color of a star. The color index employed in this figure expresses the difference in magnitude ($b − v$) measured at two wavelengths, those of blue and yellow light. The letter d denotes a double star, m a star whose spectrum shows prominent metals.

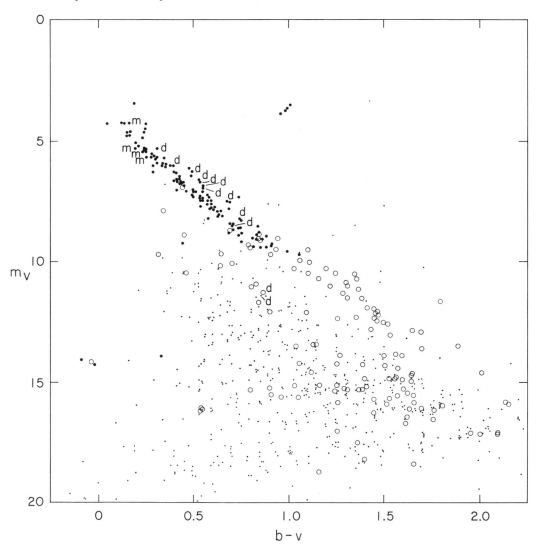

ones. Above and to the right of the main sequence lie the subgiant, giant, and supergiant stars; below and to the left are the white dwarfs (see fig. 1.4).

It is possible to build up a composite color–magnitude diagram from a series of well-observed clusters, as illustrated in figure 2.5. Each one contributes a part, not always the same part, of the diagonal backbone identified as the main sequence. This figure has been built up from a few open clusters only. A similar picture unites the globular clusters, but their main sequences represent only the lower part of the diagonal strip and do not quite coincide with the composite main sequences for the open clusters.

We note that almost all the stars in figure 2.4 lie on or near to the right of the diagonal formed by the main sequence. The exceptions, the white dwarfs, are of great interest. Many of the brightest stars in the Hyades (the subgiants) swing off slightly to the right of the diagonal strip, and four lie very far from it (giant stars). All clusters show this effect to some extent, and it furnishes a sensitive test for their ages.

An appreciable number of stars in the Hyades lie on a line above and roughly parallel to the lower boundary of the strip. Many of these stars are known to be double (they are either visual or spectroscopic binaries). If the members of a pair of stars so close together that they are not seen separately were exactly equal in brightness, each would be three quarters of a stellar magnitude fainter than their combined light. It seems likely that binaries account for most of the spread in the observed main sequence of the cluster. It is not surprising that so many stars are thus affected—at least half the known stars in our own neighborhood are known to be double.

Luminosity and color lead at once to two other important properties of the stars: effective temperature* and size. We find that the observed main-sequence stars in the cluster range in temperature from about 12,000 degrees to about 3000 degrees Kelvin and in diameter from about 2.25 to about 0.45 times that of the sun. The giant stars in the Hyades have surface temperatures of about 5000 degrees and have about twenty-five times the sun's diameter; the white dwarfs, 12,000 degrees or less and have only one or two hundredths of the sun's diameter.

---

* The effective temperature of a star is the temperature at which a "black body" (a perfect radiator) of the same surface area as the star would radiate the same total energy. The stars are not perfect radiators, largely on account of the presence of absorbing atoms and molecules in their surfaces. But for all stars except the very coolest, the effective temperature is not far from the temperature indicated by the physical condition of the atoms and molecules at their surfaces.

Figure 2.5. Course of the color–magnitude array for a number of well-known open clusters of different ages. From top to bottom: NGC 6231, κ Crucis (NGC 4755), h, χ Persei (NGC 864, 886), Pleiades, Hyades, NGC 752, Messier 67 (NGC 2682), and NGC 188.

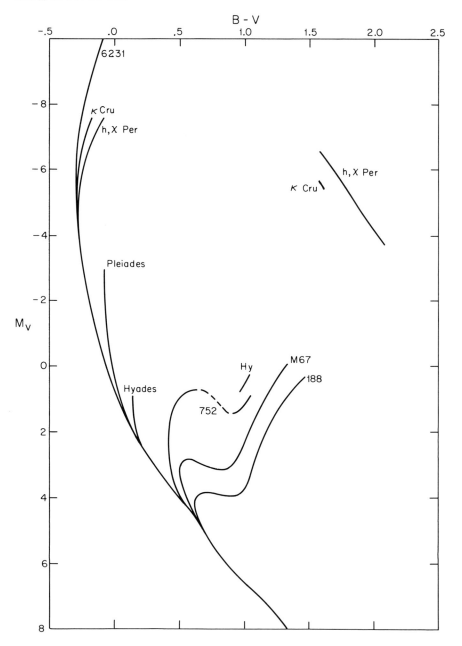

Figure 2.6.   Light curve of the eclipsing variable V 471 Tauri, a probable member of the Hyades. The system consists of a cool main-sequence star and a white dwarf. But unlike many such systems, it is not cataclysmic, although the form of the curve is variable, more so in yellow, less so in ultraviolet light. The diagram shows the ultraviolet variations. (From B. Cester and M. Pucillo, *Astronomy and Astrophysics*, 46 [1976]: 198.)

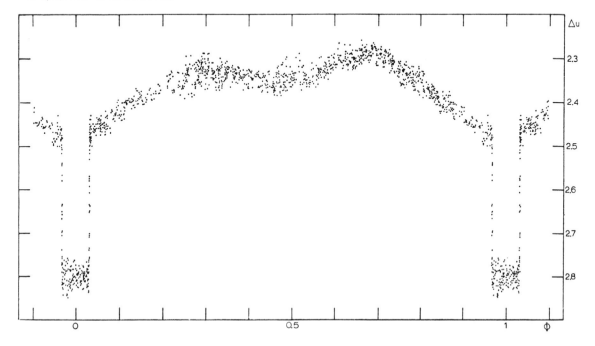

Variable stars are scarce in the Hyades. None of the double stars that were noted in the color–magnitude diagram is so oriented as to perform eclipses. One eclipsing star, V 471 Tauri, is a probable member of the cluster, and it is very odd, consisting of a white dwarf and a cool main-sequence star (fig. 2.6). Usually such pairs are cataclysmic variables or potential novae. But this one does not seem to be subject to the characteristic flickering of brightness. There seem to be a very few of the spasmodically variable T Tauri stars among the faint members of the cluster. By contrast, we shall find such stars numerous in the Pleiades and very numerous indeed in some other clusters and associations.

Another fundamental property of a star is its mass. It is a happy circumstance that the Hyades cluster contains so many double stars, for double stars are the only direct key to stellar masses. Five such systems in the Hyades show a well-defined relationship between luminosity and mass: the most massive are the most luminous. This represents the fundamental mass–luminosity relation for main-sequence stars.

To summarize: The Hyades, a typical open cluster, comprises about four hundred stars, most of them dwarf stars that conform reasonably well to the usual main sequence represented in open clusters. There are four red giants and several white dwarfs. The double stars show a well-defined mass–luminosity relation. The cluster is confined within a radius of about 14 parsecs, and most of the members are within 4 parsecs of the center, so the cluster is fairly concentrated.

The Hyades are surrounded by a swarm of fellow travelers, a moving group of stars whose motion, measured in three directions, is nearly the same as that of the cluster. They are spread throughout a much larger volume of space, about 100 parsecs in radius, so that we actually lie within its boundaries, though our sun is not a member of it. It has at least as many members as the Hyades cluster, but because membership can be assigned only if the motion has been measured, the census is less likely to be complete, especially for the fainter stars in the southern hemisphere.

The makeup of the Hyades group, as this larger body of stars is called, is very like that of the Hyades cluster. There are several cool giant stars, but most members are main-sequence stars or stars that are near to the main sequence. The hottest stars in the group are of slightly higher temperature and are slightly more luminous than any in the cluster. Like the Hyades, the group includes a large proportion of binary stars. Cluster and group are evidently "birds of a feather."

Figure 2.7 shows the structure of the Hyades group as deduced from the spectra and luminosities of the members. Unlike the diagram for the Hyades cluster, it includes only stars brighter than apparent visual magnitude 6.5. The large size of the group is apparent. We also note that the cluster does not lie in the middle of the group but is near to one edge. This may partly be a result of selection, but by including only the apparently brightest stars I have attempted to minimize this effect.

The relation between the cluster and the group is probably to be understood in terms of slight deviations from common motion caused by the perturbations that the original cluster has undergone in the course of its circulation within the Galaxy. It is tolerably certain the cluster and group had a common origin, that the cluster is gradually dissolving into the group, and that together they represent the residuum of a larger complex.

A further observation fortifies this view. Careful analysis of the spectra of stars in the Hyades cluster shows that they are not identical with the sun in composition. They are about twice as rich in metals, relative to hydrogen. The same seems to be true of the Hyades group, which contains a number of metal-rich stars.

The double stars in the Hyades cluster show a mass–luminosity rela-

Figure 2.7. Stars assigned to the Hyades group, viewed from an angle of 30°
above the galactic plane. The stippled area represents the plane, marked off at intervals of 10 parsecs from the sun, which is at the center. The galactic center is to
the left of the diagram, the anticenter to the right. Stars that lie above the plane
are shown by dots; those below the plane by circles. The vertical distance of each
star from the plane is shown by a line. The broken circle shows the dimensions of
the Hyades cluster, which lies below the plane. In an attempt to avoid the effects
of selection, I have included only the brighter stars that have been assigned to the
Hyades group.

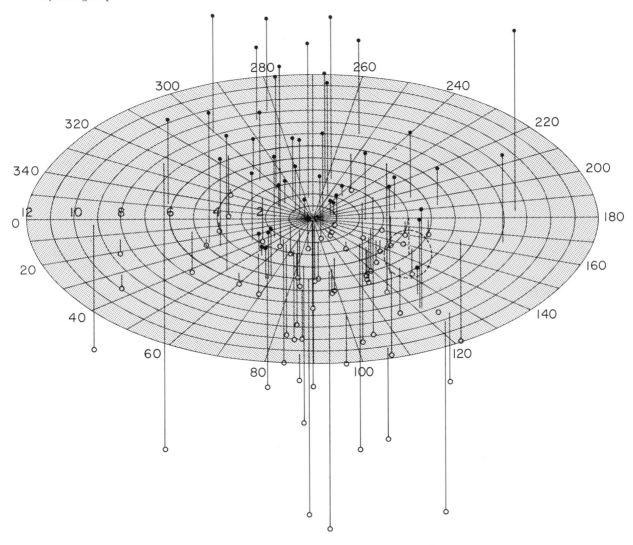

Figure 2.8. Left: relation between absolute visual magnitude, M_v, and the logarithm of the mass in units of the sun's mass. Open circles denote members of the Hyades cluster; crosses denote members of the Hyades group. Right: relation between absolute visual magnitude, M_v, and the logarithm of the mass in units of the sun's mass. Dots denote members of the Ursa Major group; crosses denote members of the moving group that includes Sirius. In both diagrams the circled dot represents the sun. The actual difference between the relations shown in the two diagrams depends critically on the distances, which are still being refined. Recent revisions tend to reduce the difference, though not to remove it entirely. (From Olin J. Eggen, *Annual Review of Astronomy and Astrophysics*, 5 [1967]: 113, 114.)

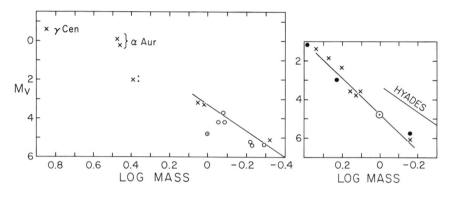

tion that may differ slightly from the relation shown by stars of sunlike composition (fig. 2.8). The double stars of the Hyades group show a mass–luminosity relation similar to that shown by the stars of the cluster, again not quite identical to that deduced for sunlike stars, typified by the Sirius group. Probably the difference in composition is associated with the slight difference in the mass–luminosity relation. It emphasizes the fact that in order to specify the properties of a star we need more than luminosity, diameter, temperature, and mass: the additional parameter is chemical composition.

The Hyades cluster and the Hyades group, then, represent stars that have been associated in space for a long time. It is hard to imagine that a haphazard collection of stars could arrive at so neat a concatenation of fundamental properties and possess a common motion in space. The group has been in existence in the form of stars for about a hundred million years; the Hyades cluster's properties suggest that it is slightly older than the youngest members of the Hyades group—a not uncommon relationship between an association and a cluster that forms its nucleus. The common motion and common composition imply a common origin in space from the same original mass of material.

We cannot say, however, that all the member stars originated at exactly the same time. In fact, the group contains stars that seem younger than those of the cluster. The stars in question are brighter and bluer than the brightest main-sequence stars in the Hyades cluster. Similar apparently anomalous stars are found in other clusters. Star formation can have been going on for an appreciable fraction of the lifetime of the cluster, although the clear-cut relationship between luminosity and color sets limits to this fraction. A comparison with other clusters that display different inner relationships can throw light on this question.

The Hyades will appear again in chapter 9. As the nearest and best-studied stellar community, it serves not only as a specimen but as a warning of the limitations to our detailed knowledge of clusters.

The Food of the Stars

CHAPTER 3

A glance at the Hyades reminds us that the stars are not all alike. Even in this close-knit stellar community, the most luminous star is more than a million times as bright as the faintest; the bluest has about four times the surface temperature of the reddest. The variety is even more evident when the spectra of the stars are examined (fig. 3.1). The bluest are dominated by absorption lines of hydrogen; there are also many weaker metallic lines (iron, titanium, magnesium) and the strong "H and K" lines of ionized calcium. The fainter, redder stars show a progressive weakening of the hydrogen lines and a strengthening of the lines of neutral and ionized metallic atoms. In the spectra of the very faintest stars the neutral metals are very prominent, and absorption bands of metallic oxides appear. Midway down the main sequence the stars appear to resemble the sun, though metallic atoms are rather more conspicuous. The four bright red outliers have strong metallic spectra and weak hydrogen; the faint blue outliers are dominated by very broad absorption lines of hydrogen.*

---

* Like the clusters and galaxies, the spectra of the stars have received unromantic names, despite the interesting personalities they reveal. In this case the notation is alphabetical, decimalized, and subdivided by roman numerals. It can be summarized here only in the most general terms, for there are few subjects in which the details are more intricate and specialized.

The modern system is decimal, each letter class having potentially ten subdivisions that grade continuously one into another. Every fifth subdivision is given in table 3.1. Luminosity class is denoted by roman numerals: I for supergiants, II for bright giants, III for giants, IV for subgiants, V for main-sequence stars, and VI for subdwarfs. White dwarfs have a special notation. The hottest stars (O and B) are dominated by highly ionized spectra in which helium is especially prominent. Class A is dominated by hydrogen, but the lines of ionized metallic atoms are also present. In classes F, G, and K the metallic atoms are progressively more prominent, hydrogen less so; lines of neutral atoms intensify progressively at the expense of ionized lines. Spectra of classes M and S are dominated by metallic oxides

Figure 3.1.   Spectra of stars in the Hyades. (Photograph by Harvard Observatory.)

and other molecular species; those of class C (which has superseded class N) characteristically show the spectra of carbon compounds. Absorption lines are the rule, but emission lines occur as peculiarities everywhere in the sequence, and in the exceptional, hot Wolf-Rayet stars they dominate the spectrum.

Effective temperatures are deduced, on the assumption of black-body radiation, from the energy emitted at the star's surface. They follow the same sequence as, but are not identical with, the temperatures deduced from the observed line spectra; the spectra of stars are produced by the passage of their radiation through the layers of an atmosphere that is often deep and may be violently disturbed and inhomogeneous.

Table 3.1. Absolute visual magnitude (according to luminosity class) and effective temperature for selected spectral classes. Effective temperatures are tabulated for the main-sequence (luminosity class V) representatives of the spectral classes listed. Effective temperatures for stars of luminosity classes I (supergiants), II (bright giants), III (giants), and IV (subgiants) are in general somewhat lower: for example, 5,260°, 4,720°, and 3,500° Kelvin for G5 III, K0 III, and K5 III stars, respectively. Colons denote uncertain values; dashes, absence of information.

Spectral class	Absolute visual magnitude					Effective temperature (°K)
	V	IV	III	II	I	
O5	−5.1	—	—	—	—	35,000
B0	−4.1	−4.6	−5.0	−5.6	−6.2 to −7.0	30,000
B5	−1.1	−1.6	−2.2	−3.7	−5.7 to −7.0	16,400
A0	+0.6	0	−0.6	−2.8	−4.9 to −7.0	10,800
A5	+2.1	+1.2	+0.3	−2.1	−4.5 to −7.0	8,620
F0	+2.6	+1.7	+0.6	−2.0	−4.5 to −7.0	7,240
F5	+3.4	+2.1	+0.7	−2.0	−4.5 to −7.0	6,540
G0	+4.4	+2.8	+0.6	−2.0	−4.5 to −7.0	5,920
G5	+5.2	+3.2	+0.3	−2.1	−4.5 to −7.0	5,610
K0	+5.9	+3.2	+0.2	−2.1	−4.5 to −7.0	5,240
K5	+8.0	—	−0.3	−2.3:	−4.5 to −7.0	3,970
M0	+9.2	—	−0.4	−2.4:	−4.5 to −7.0	3,600
M5	+12.3	—	—	—	—	2,800

This variety of spectra is not, of course, necessarily evidence of variety in chemical composition. When stellar spectra were first observed, they were thought to furnish evidence of gross differences of chemical composition, from hot helium stars to cool metallic stars. This early simplistic view has long been superseded by a complex and sophisticated theory of stellar atmospheres which interprets the major differences in terms of temperature, density, and atmospheric structure.

In broad terms, the composition of nearly all stars is dominated by hydrogen, with lesser contributions from atoms of greater atomic weight. Most of the known atomic species are represented in stellar spectra, and, though many details still remain unidentified, there is no reason to believe that stellar spectra contain the lines of atoms unknown on earth. At first glance the chemical composition of the universe is surprisingly uniform.

This is not to say that all stars are completely identical in composition. Careful study, abetted by sophisticated theoretical models of stellar atmospheres, reveals real differences. The Hyades stars contain a higher proportion of metallic atoms than the sun. Some anomalies can be understood in terms of special conditions in the stars' atmospheres, but not all. A few exceptional stars—v Sagittarii* and R Coronae Borealis, for example—contain very little hydrogen; their composition is dominated by helium. Others show evidence of exceptional quantities of mercury, and even of the short-lived unstable atom of technetium.

Conclusions such as these are drawn from a detailed study of the spectra of stellar surfaces, the only parts that can be directly analyzed. There is reason to think that stellar interiors differ likewise in composition, that these differences are progressive and inherent in stellar development, and that they are reflected in the surface properties, behavior, and life history of the stars.

The array of surface properties shown by the Hyades, a relatively coeval group, is but one example of the pattern shown by the many hundreds of clusters and associations. Not all resemble the Hyades. Some contain stars that are far hotter and brighter. Some contain many red outliers, like the four giants in the Hyades, some none at all. Many contain variable stars; in fact there is virtually no type of variable star—great as is their variety—that cannot be found in some cluster or association. Hence, the variable stars, too, can be assigned their place in the pattern. Their behavior is part of a natural sequence of events.

How does it happen that the Hyades, confined within a sphere of about

---

* Most of the brighter stars have names that identify them within one of the eighty-eight constellations. The brighter naked-eye stars (such as α Andromedae) were designated within each constellation by Greek letters, roughly in order of brightness, by Johann Bayer in the seventeenth century. The brighter stars in each constellation were numbered serially by John Flamsteed a little later (for example, 12 Lacertae). In addition, many of the brightest stars have names (such as Sirius) that go back to remote antiquity. Where such a historical name exists, I have tended to make use of it, adding also the Bayer letter for identification on star charts.

Many stars have a variety of names if they have been studied from several points of view—position, spectrum, motion, or variability, for instance. An example is furnished by Pleione, one of the Pleiades, which also bears the names 28 Tauri (from the Flamsteed number), B.D. +23° 558 (from the entry in the first modern star catalogue, the *Bonn Durchmusterung*, which means that the star is number 558 in the zone of declination that extends from +23° to +24° as observed in 1855, the date at which the stars were catalogued at Bonn Observatory), HD 23862 (from the *Henry Draper Catalogue*, where its spectrum was classified), GC 6841 (from its number in the *General Catalogue of Proper Motions*), and finally BU Tauri, the designation given to it as a known variable star.

Variable stars, too, are arranged by constellation. When few variable stars were known, it was thought convenient to designate them by letter, beginning with R, within each constellation. As discoveries progressed beyond expectation, the sequence R, S, T, . . . Z proved inadequate, and the series was started again: RR, RS, . . . SS, ST, . . . ZZ, AA, AB, . . . AZ, BB, BC, . . . QZ. As variable stars piled up, many constellations reached QZ, and instead of proceeding to three-letter names, the "earthly godfathers of Heaven's lights" continued with V 334, V 335 . . . , a series that can go on indefinitely.

10 parsecs in radius and moving together through space, differ so greatly among themselves in luminosity, diameter, and temperature? They must have been together from the first (whatever that means), for the very pattern of their properties makes it wildly improbable that they have come together by accident. They must have been formed from the same body of material, not perhaps at exactly the same time but within a limited interval. What conditions have dictated their current variety?

The clue is to be found in the relation between luminosity and mass. The color–luminosity relation of the Hyades is somewhat blurred as a result of the happy circumstance that the cluster contains a number of double stars—happy because the masses of these double stars can be determined. From the Hyades alone we can conclude that the more massive a star on the main sequence, the more luminous it is. At once the possibility is suggested that the stars differ in surface properties because they differ in mass.

The work of determining stellar masses is as laborious and exacting as that of sorting out the members of a moving cluster. More so, indeed, for after the apparent orbit of a visual binary has been determined—the fruit of years, perhaps centuries, of measurement—a sophisticated application of celestial mechanics is needed for calculating the masses of the components. The stars whose masses have been directly determined still number only a few dozen. They are shown in figure 3.2.

The known masses of main-sequence stars fortify and extend the conclusion at which the double stars in the Hyades could only hint. There is a clear-cut relation between mass and luminosity for main-sequence stars: the most massive stars are the brightest. Moreover, luminosity is not directly proportional to mass, but to between the square and the cube of the mass.[*]

Such preliminaries may seem to furnish a devious approach to the subject of this chapter, the theory of stellar evolution (or rather, stellar development, because the process concerns the history of an individual star rather than the transmission of characteristics through inheritance). But evidence such as that gained from the Hyades has pointed the way to the current theory of stellar development.

Some assumptions—we might almost call them axioms—are inevitable. The theory assumes that the members of a cluster originated together from the same material. Another assumption (justified by its suc-

---

[*] This is the theoretical relationship for main-sequence stars, to which alone it is strictly applicable. Empirically, the exponent of the relationship (which connects total radiated energy with mass) seems to be somewhat smaller for stars with masses less than that of the sun.

Figure 3.2. Mass–luminosity relation for main-sequence stars. Visual binaries of high weight are indicated by large, heavy dots; those of low weight, by small, light dots. Eclipsing binaries are indicated by squares. The solid line represents the adopted mass-luminosity relation for late-type dwarf stars. (From D. L. Harris III, K. A. Strand, and C. E. Worley in *Basic Astronomical Data*, ed. K. A. Strand [Chicago: University of Chicago Press, 1963], p. 285.)

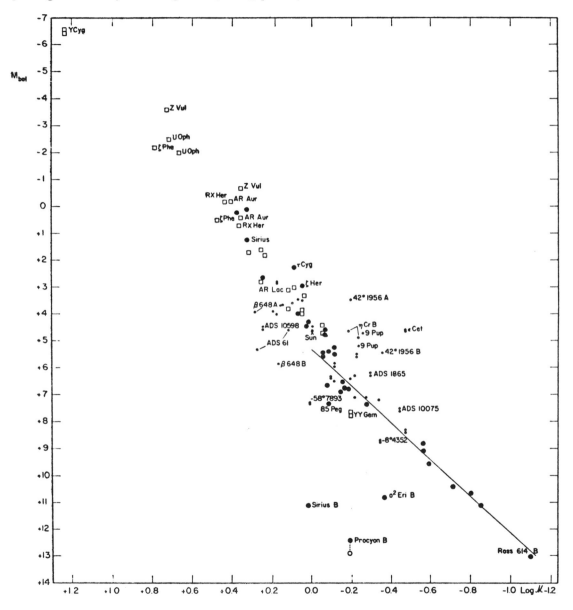

cessful application) concerns the source of stellar energy. There was a time when the stars were thought to be drawing on gravitational energy to supply thieir radiation, but for the better part of a century this source has been recognized as hopelessly inadequate except perhaps during the earliest and last stages of development. Slowly it became evident that the stars shine for most of their lifetimes by devouring their own substance. To the modern world the role of nuclear energy has become a commonplace, and it is no longer a mere assumption that thermonuclear energy is the source of the radiation of the stars, at least for the major part of their careers.

If thermonuclear energy is to be the main source of stellar energy, a few simple conditions must be fulfilled. The nuclear fuel must be present in the stars in adequate supply, the conditions for the release of its energy must exist, the reactions being necessarily exothermal (accompanied by release of energy). Hydrogen, commonest of all elements, is in fact the primary fuel. From quite elementary considerations it is found that if the masses, sizes, and densities of the stars are what they are known to be, the central temperatures must be of the order of ten million degrees. Thermonuclear release of energy in stellar interiors under these conditions can be predicted, and the current theory of stellar development is the outcome of these calculations. A star consumes its hydrogen and produces atomic nuclei of progressively greater atomic weight; these in turn are consumed under the consequently changing conditions.

A detailed description of the thermonuclear reactions that can take place in stellar interiors is outside the scope of this book. I confine myself to a brief outline. The simplest reaction builds up hydrogen into helium, and in this reaction the carbon nucleus acts as a catalyst. Hydrogen burning is at first confined to the core of the star. As the hydrogen there is gradually exhausted, the reaction spreads to a growing shell surrounding the core. When the available hydrogen has been used up, rising central temperature triggers the next step. Helium, the product of the hydrogen burning, now becomes the fuel. Carbon, nitrogen, oxygen, neon, magnesium, silicon, and heavier nuclei are successively produced and successively consumed. Each reaction is highly sensitive to temperature, which in turn is affected by changes in density. The structure of the whole star changes progressively in response to changing conditions, core burning being followed by shell burning.

A very important factor is of course the original chemical composition of the material of which the star is built. If it contained the light nuclei of lithium, beryllium, and boron, these would have fallen victim to rising central temperature even before the onset of hydrogen burning. These elements are indeed very rare in stellar atmospheres, and their presence in some stars suggests extreme youth, a phase before the principal sources of energy are tapped.

The calculation of the course of stellar development from the known physical properties of the relevant atomic nuclei is one of the most spectacular feats of astrophysical theory, aided by modern computer techniques. Figure 3.3 shows the predicted history of a star of five times the sun's mass, starting from its arrival on the main sequence, the point at which hydrogen burning is thought to begin (known as the *zero-age main sequence*). While enough detail is shown to emphasize the great complexity of the problem, no account has been taken of the possible effects of rotation (and many stars are observed to be rotating rapidly) or of the loss of mass which is observed to occur when a star becomes a red giant, a process that must modify the track of development. These calculations are made for a particular adopted composition: 71 percent hydrogen, 27 percent helium, 2 percent metallic atoms. Small differences of composition would modify the tracks numerically but would probably not alter their general form.

Theory predicts the rate of development as well as its course. Starting with its arrival on the main sequence (point 1 on fig. 3.3), the star of five times the sun's mass runs through its career as far as the second giant branch (point 14) in less than a hundred million years. Most of this time is spent on the main sequence or near it: some stages are relatively brief, such as the red-giant stage (point 6), less than a hundredth of the main-sequence stage. Table 3.2 gives a summary of the course of development and the times taken by the various processes.

A star must undergo great changes in its development simply because it starts out with a limited supply of the fuel necessary to its sustenance. The internal changes have only been inferred, but the surface changes that are their consequence can be matched with the properties of actual stars. When on the main sequence, a star of five times the sun's mass is nearly a thousand times as bright as the sun and comparable to a fairly hot star such as ν Andromedae, whose temperature is about 20,000 degrees. By the time it reaches the red-giant stage it has become a cool red giant, something like α Herculis (Ras Algethi), with a temperature of about 3500 degrees (a star that is observed to be losing mass from its surface). It has passed rapidly through stages at which its condition predisposes it to undergo bodily pulsations, such as are observed for the Cepheid variables. The later adventures of the star beyond the second giant branch are not covered by figure 3.3 and have yet to be described.

The picture is amplified when the course of development is calculated

Figure 3.3   The path of a metal-rich star of five solar masses (5M⊙) in the Hertz-
sprung-Russell diagram. Luminosity is in solar units ($L\odot = 3.86 \times 10^{33}$ erg/sec),
and surface temperature (T_e) is in degrees Kelvin (°K). Traversal times between
labeled points are given in years. (From Icko Iben, Jr., *Annual Review of Astronomy and Astrophysics*, 5 [1967]: 573.)

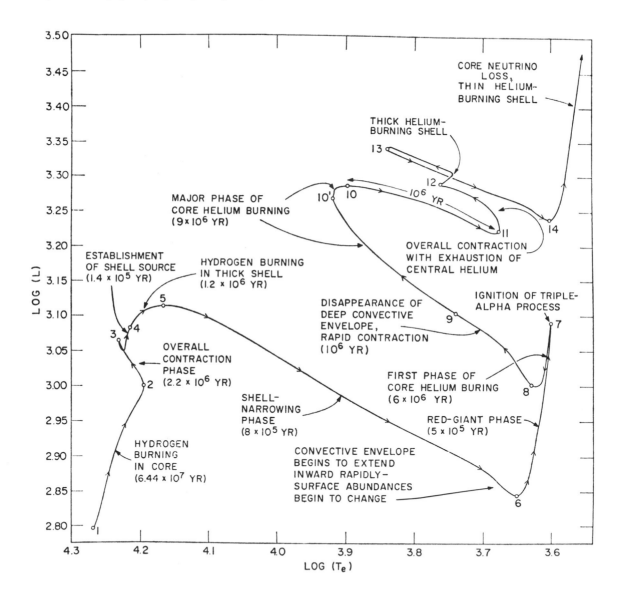

Table 3.2.   Development of a star of five solar masses. Colons denote uncertain values.

Point on figure 3.3	Phase	Duration of phase (millions of years)	Time from zero-age main sequence (millions of years)
1			
	Hydrogen burning in core: main sequence	64.40	64.40
2			
	Overall contraction	2.20	66.60
3			
	Shell source established	0.14	66.74
4			
	Hydrogen burning in thick shell	1.20	67.94
5			
	Shell narrows; convective envelope begins to extend inward	0.80	68.74
6			
	Red-giant phase	0.50	69.24
7	(Ignition of triple alpha process)		
	First phase of core helium burning	6.00	75.24
8			
	Disappearance of deep convective envelope; rapid contraction	1.00	76.24
9			
	Major phase of core helium burning	9.00	85.24
10			
		1.00	86.24
11			
	Overall contraction with exhaustion of central helium	2.5:	88.74:
12			
	Thick helium-burning shell		
13			
14			
	Thin helium-burning shell, core neutrino loss; ascent to giant branch		

for stars of similar composition with a variety of masses. In figure 3.4 such tracks are shown for stars that range from fifteen times the sun's mass to a quarter of the sun's mass. The more massive the star, the more luminous it is at every stage and the higher the temperature at which it leaves the main sequence. The tracks show a family resemblance, but they are not parallel.

Table 3.3. The cumulative time elapsed at the end of each phase of development for stars of various solar masses, measured from the zero-age main sequence. The calculations on which the table is based refer to an adopted composition of hydrogen (71% by weight), helium (27%), and metals (2%). The calculations for less massive stars are not carried beyond point 6.

Point on figure 3.4 at which phase ends	Time from zero-age main sequence for various solar masses (millions of years)							
	15.0	9.0	5.0	3.0	2.25	1.5	1.25	1.0
2	10.100	21.440	65.470	221.200	480.20	1,553.0	2,803.0	7,000
3	10.327	22.045	67.643	231.620	496.67	1,634.0	2,985.4	9,000
4		22.136	69.015	241.950	533.63	1,974.9	4,030.4	10,200
5	10.403	22.284	69.768	246.456	546.73	2,079.8	4,176.7	10,357
6		22.350	70.254	250.694	585.02	2,279.8:	4,576.7:	11,357
7	11.120	22.840	76.304	275.794				
8	11.740	23.794	77.324					
9	11.930	27.074	86.324	316.594				
10	11.965	27.229	87.254	322.594				

Adapted from Icko Iben, Jr., *Annual Reviews of Astronomy and Astrophysics*, 5 (1967): 585.

Table 3.3 shows the cumulative time elapsed at the end of each phase of stellar development for the solar masses shown in figure 3.4. A very significant result of the calculations is that the more massive the star, the shorter is the time it spends on the main sequence and the faster it runs through the later stages. At fifteen solar masses, the main-sequence stage lasts almost ten million years, in contrast to nearly ten thousand million years for a star of one solar mass. Stars of at least fifty solar masses are known, and their main-sequence lifetimes must be very brief indeed— less than a million years. And at one tenth of a solar mass the main-sequence lifetime would be much greater than the probable life of the Galaxy. Although stars of greatest mass run their courses most rapidly, all stars spend most of their time on or near the main sequence.

Figure 3.4 shows the properties of stars of different masses after various amounts of time have elapsed. But what we actually observe when we study a cluster are stars of different masses after the *same* amount of time has elapsed (or nearly so)—a so-called *isochrone*. The stars of a given cluster define the isochrone that corresponds to the age of the cluster; it is the pattern presented by a number of stars of different masses that set off from the zero-age main sequence at nearly the same time.

The isochrones bear a close relation to the tracks of development. In figure 3.4 and table 3.3, for every mass represented, the whole course of

Figure 3.4. Tracks of development for stars of various masses. Left: total luminosity (light of all wavelengths) in terms of the luminosity of the sun is plotted against effective temperature; both scales are logarithmic (see fig. 3.3 for explanation). The theoretical tracks follow the numbered points; the corresponding masses, in terms of the sun's, are shown at the left-hand ends of the tracks. The dotted line shows the zero-age main sequence.

Right: the same data have been transformed into the quantities used in the study of clusters: absolute visual magnitude, M_v, is plotted against color index $(B - V)$. The fact that the tracks are not the same shape in the two diagrams is a result of the fact that the B and V magnitudes, and the absolute V magnitude M_v, refer to limited ranges of wavelength. (Adapted from Icko Iben, Jr., *Annual Review of Astronomy and Astrophysics*, 5 [1967]: 585.)

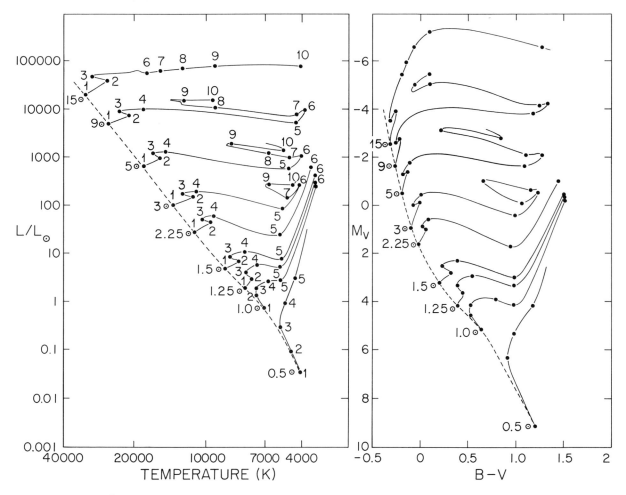

development shown in the diagram has long been over before the star of next lowest mass has reached even point 2. So stars of equal ages will present patterns that lie, for each decrease in mass, nearer and nearer to the point of departure from the main sequence. After point 4 the stars begin to pass rapidly across the color–magnitude domain, and the fraction of their lifetime occupied by this phase is so small that the isochrones are nearly of the same form as the tracks of development. For the series of masses shown in figure 3.4, the isochrones for any one mass always lie between the development track for that mass and the track for the next lower mass represented. For low masses the tracks converge. Thus, in the isochrone for the most massive stars, the luminous red stars are visually fainter than the brightest blue stars; for stars of intermediate mass the absolute visual magnitudes are about equal; and for the least massive, the red stars are visually the most luminous. We shall see that the clusters bear out this prediction.

The isochrones provide a means of assigning an age to a group of coeval stars: the point at which the color–magnitude diagram swings away from the main sequence reveals the upper limit of mass for a star that is still on the main sequence. And this upper limit, in turn, reveals the age of the group, since duration of the main-sequence stage is uniquely related to mass. Thus, groups with very luminous (and hence massive) stars still on the main sequence are young; those whose brightest main-sequence stars are of low luminosity are old.

In speaking of clusters and associations we shall follow this procedure for assigning ages. But some qualification must be made. The basic calculation involved a chosen initial composition, and initial composition may differ from group to group, or even conceivably within a group. We have already noted that the Hyades are richer in metals than the sun, and such a difference may well modify the tracks, and even more probably the times of traversal. Second, the procedure implies that the stars of a given group all set off from the zero-age main sequence at the same time, which may not be strictly true. The fact that many color–magnitude diagrams such as that of the Hyades are as neat as they are suggests that the stars must have set off within a small time interval compared to the total age of the group. But there is growing evidence that all the stars in a cluster did not form simultaneously. The scatter of the color–magnitude diagrams of the youngest clusters tends to be especially large. A formation span of a million years would obviously blur them much more than it would do in the case of a cluster whose main-sequence stars were less massive and hence developing more slowly.

Here we touch on a very sensitive question. There is no clear evidence as to what determines the distribution of masses during the formation of a cluster. What is clear from observation is that in any one group the most luminous (hence presumably the most massive) stars are the rarest, and that less luminous stars are progressively more frequent. This distribution of luminosities is also true for stars at large—"field stars" that are not members of clusters, or at least are not now known to be associated with clusters.

We shall find when we come to consider the oldest open clusters, and still more so when we meet the globular clusters, that they do not contain any very luminous (and presumably very massive) stars. Did they ever contain the counterparts of the brightest stars of very young clusters, and if so, what has happened to these stars? Unless the "stellar birth function" has changed with time, they must have undergone a drastic change. It is not without significance that the known stars of very high mass are luminous blue stars: once they have left the main sequence, such stars cannot last long. The giant eclipsing star UW Canis Majoris (29 Canis Majoris, another star of many names), and the second magnitude ζ Puppis, one of the hottest known stars, are of very high mass, at least thirty times the sun's. We do not know what the future holds for them.

We have taken the main sequence as the starting point. But how do stars reach the main sequence? It must be supposed that they are formed from aggregations of matter, presumably by a process of contraction. Many embryonic stars are invisible, deeply embedded in thick obscuring material. It can be supposed that as they contract, their internal temperature rises until it activates the thermonuclear reactions that characterize main-sequence stars. Theory suggests that the process lasts roughly 10 percent of the time spent on the main sequence, so the age deduced from an isochrone must be increased by about 10 percent if we wish to reckon from the time when a star assumed a definite identity—a vague conception at best. In any case, when the career of a star up to the giant stage is considered, the greater part of the time is spent on or near the main sequence.

The results pictured in figure 3.4 constitute a theoretical mass–luminosity relation for main-sequence stars with the particular composition used in the calculations. They reproduce the observed mass–luminosity relation reasonably well. It is evident that stars that have left the main sequence will not necessarily be expected to conform to the mass–luminosity relation, and in fact most of the few such stars whose masses have been determined do not do so.

We have followed the adventures of a star from the pre-main-sequence stage to the tip of the giant branch. There is not a steady one-directional development. The track alters, bends back on itself, and may even execute loops. Each section of the track is associated with a definite stage or type of behavior, and each change in direction goes with a change in the stellar digestive process. Pre-main-sequence stars are gravitationally fed; main-sequence stars are devouring hydrogen. The site of hydrogen burning and the internal structure alter as the star matures. By the time it reaches the red-giant stage the core of the star is ready for a diet of helium, and as the site of helium burning moves outward from the core into a shell the star can become a Cepheid variable. With each change of diet a star alters its outward character.

If all stars passed through these stages at the same rate, the predicted pattern of stellar development represented by the isochrones would move bodily in the direction of the redder, cooler stars. But the loci of the stages reached after the same interval by stars of various masses emphasize that this is not so. Stars of greatest mass develop fastest. The point of departure from the main sequence, when the hydrogen in the core has been digested, comes earlier for massive stars and can be used as a measure of the age of a group. As we survey the color–magnitude diagrams of open and globular clusters, we shall see a series of isochrones corresponding to times that range from less than ten million to over a thousand million years.

The "time horizon" of our galaxy is something over ten thousand million years. We cannot identify a star or a cluster that is demonstrably older than that. In this interval stars with less than a tenth of the sun's mass will not have reached the turning point; they are still consuming the hydrogen in their cores. Stars of ten times the sun's mass will by now have left the main sequence and passed the giant tip, to an ultimate fate that carries them beyond the bounds of our diagram. If we find any such stars still on the main sequence, they cannot be so old. For an isolated star this merely sets an upper limit to age. But when a star is a member of a group whose color–luminosity pattern is known, age can be deduced from the point at which stars are just deviating from the main sequence.

First the Infant—The Pre-Main-Sequence Stage

From a survey of the Hyades and an excursion into ideas that relate the processes of stellar nutrition to the observed properties of the stars, we have reached an empirical picture of the main sequence and a general theoretical notion of its significance. It marks the starting point of the thermonuclear epoch of the stellar lifetime. The steps by which a star reaches the main sequence have only been roughly sketched.

For many reasons the sketch must remain rough. Beginnings are necessarily elusive. Given a definite mix of atomic nuclei and elementary particles and a definite start at material aggregation with a variety of masses, it is not hard to get the stars rolling. But the genesis of at least some atomic species must be assumed; it cannot be directly observed on the cosmic scale. It is difficult to start star building with pure hydrogen.

Neither can I sketch the origin of a stellar system, which is the province of the cosmologist and far transcends the kind of information my parochial approach can offer. The existence of the Galaxy, and of the interstellar medium and environment from which the stars are born, are as axiomatic for my purposes as the existence of the atomic species. My illustrations are taken, and my conclusions are drawn, from things that can be seen and examined in our immediate vicinity. Only rarely shall we step outside the galactic system; only reluctantly shall we touch on the unfamiliar and exotic.

The origin of stars, then, remains elusive. Probably they do not form in isloation but originate in groups from the clumping together of pre-stellar material. Rather than speculate on *how* stars are formed, it may be profitable to examine *where*. Perhaps stars that are found only in groups may provide clues to stellar formation.

Figure 4.1.   Apparent distribution of pre-main-sequence variables over the face of the sky, which is shown in an equal-area projection, with the galactic anticenter in the middle. The tendency to occur in groups is evident. The dense groups near the middle include the Orion and Taurus nebulosities, the Pleiades, and the clusters in Monoceros. In these groups the variables are so numerous that it has not been possible to show them all individually. The little group near the north galactic pole is associated with the cluster Coma Berenices.

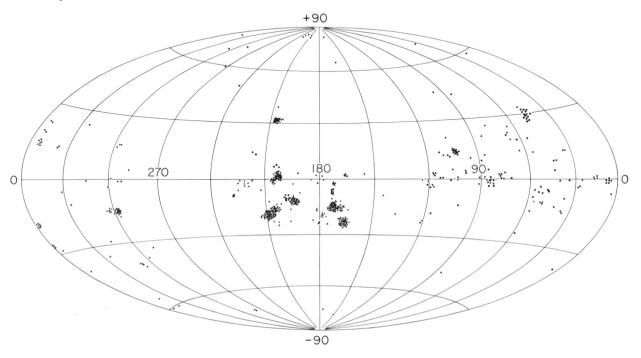

The first hint of such stellar hatcheries was noted many years ago by Alfred H. Joy, with his unerring instinct for significant facts. He found that a large number of faint, irregular, variable stars lie within dark nebulosities in the constellation Taurus, not far from the Hyades in the sky but actually at a greater distance from us. He pointed out that there was an exceptionally large number of these stars in a small volume of space and correctly surmised that they had recently been formed within the dark nebula. These so-called T Tauri stars (named after one of the first of them to be discovered) have given their name to the "T associations," which include a number of other similar groups. Several thousand such stars, or stars of related types, are now known.

Figure 4.1 shows the apparent distribution of these variable stars in the sky. Evidently they tend to be concentrated in dense groups, several of which are identified in the diagram. If we had included stars that resemble these variable stars in spectrum but are not known to vary in brightness, the groups would have been even more populous and compact. The compactness is real, for all these groups of irregular variables are close to us and are in fact small.

The picture is emphatically incomplete. It reflects the places where intensive searches for such stars have been made and represents only a small volume of space. Moreover, many of these stars are intrinsically faint, and most of them are enmeshed in dark nebulosity, which makes them hard to observe even at short distances.

There are no bright, familiar stars among these irregular variables. Representatives of most kinds of variable stars can be seen with the naked eye, but the brightest of the pre-main-sequence stars—T Tauri, RY Tauri, and RU Lupi—are ninth-magnitude stars, and most are much fainter.

That these are the very youngest stars is suggested by the fact that with extremely few exceptions they are found *only* in groups. We might expect it to follow that the groups would consist exclusively of these irregular variable stars, but this is not the case. Some of the groups contain many stars that are not known to be variable at all, though they often resemble the variables in their spectral peculiarities. Others display an array of members consisting not only of main-sequence stars but also of evolved stars that are already on their way to becoming giants.

In fact, our chart of the sites of irregular variables covers a variety of stellar communities. The Hyades, Coma Berenices, and the Pleiades include representatives, and all these groups have well-defined main sequences; the two first also contain evolved stars. The great twin clusters h and χ Persei are associated with a few, and so is the o Persei cluster (IC 348). A larger number are found in the nebulous clusters immersed in the Trifid Nebula (Messier 20) and the Lagoon Nebula (Messier 8, both shown in fig. 4.2), in NGC 6611 (fig. 4.3), and in NGC 2264 (fig. 4.4). The great Orion association swarms with them. All these clusters and associations have main-sequence backbones. The groups in Taurus, in Corona Austrina, and in Ophiuchus have no obvious main sequence. The positions of several of these clusters are given in table 4.1.

The groups just named have been mentioned roughly in order of age, beginning with the oldest, as deduced from the level reached by the main sequence. Irregular variables play but a small part in the known population of the Hyades; they preponderate in the Orion complex and are the dominant inhabitants of the groups embedded in dark nebulosity, such as the Taurus and Chameleon clouds. The region of the well-known variable star R Monocerotis, embedded in nebulosity and probably only visi-

Figure 4.2. The Trifid nebula (Messier 20) and the Lagoon Nebula (Messier 8). They lie in a region permeated by dark and bright nebulosity. Both are centers of bright nebulosity and regions of active star formation. The Trifid Nebula contains a few pre-main-sequence variables of the seventeenth and eighteenth magnitudes. The Lagoon contains forty or fifty such stars associated with a cluster of very hot, bright stars and a number of dark "globules." (Photograph by Boyden Observatory.)

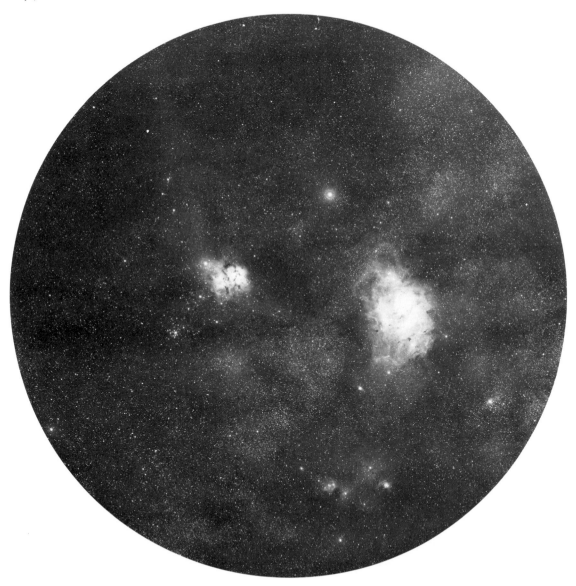

Figure 4.3. The nebulous cluster NGC 6611 (Messier 16), which contains a number of pre-main-sequence variables. (Photograph by the United States Naval Observatory.)

Figure 4.4. The nebulous cluster NGC 2264, which contains many pre-main-sequence stars. (Photograph by Harvard Observatory.)

Table 4.1.   Positions of four open clusters that contain pre-main-sequence stars.

Name	Position			
	h	m	°	′
IC 348	3	38.3	+31	56
NGC 2264	6	35.5	+ 9	59
NGC 6530	17	58.6	−24	23
NGC 6611 (M16)	18	13.2	−13	49

ble by reflected light, is shown in figure 4.5. There is little nebulosity in the Hyades, more in the Pleiades; the nebulous clusters are wreathed in clouds, both bright and dark. Finally come the groups of stars within the dark nebulae. If these groups, like the Lagoon Nebula (Messier 8), contain hot, bright stars, the high-temperature radiation would cause the nebulosity to glow; but no such stars are visible in them.

Although the dark nebulosities conceal what lies within them from optical observation, the deep interiors of the dark clouds can be probed by study in the infrared and the radio regions of the spectrum. The little

Figure 4.5.   The variable star R Monocerotis, embedded in a variable nebula. Left, blue light; right, red light. (Photographs by Harvard Observatory.)

group of variable stars in Corona Austrina, long a known T association, includes a small compact nebulosity that can be indentified as a so-called Herbig-Haro object (fig. 4.6).* These small, bright knots have bright-lined spectra of relatively low excitation and were at first regarded as possible embryo stars. Like the irregular variables, they are found in regions of dark nebulosity and they too vary in brightness. A strong infrared source of light is observed to be near (but not coincident with) this particular Herbig-Haro object, and it now seems probable that the little nebula is illuminated by an otherwise undetectable star that lies inside the dark cloud. The star seems to be an irregular variable. Similar obscured regions are associated with NGC 1333 (fig. 4.7) and the Chameleon cloud (fig. 4.8).

There is a growing body of evidence from observations of this kind that there are stars within these dark nebulosities whose light is obscured by as much as 10, 20, or 30 magnitudes—bright stars that are revealed only by their reflected light. The dense nebulosity reddens them so that they can only be detected in the infrared. Or is their redness evidence of stellar infancy at extremely low temperature?

This question cannot yet be answered unequivocally, but it seems likely that the redness is in fact the effect of obscuration. Observations made with the radio telescope of the dense, dark nebulosities have revealed high excitation of molecules such as those of cyanogen and carbon monosulphide, which points to high-temperature radiation from hot stars. It seems that in their early embryonic stages the nascent stars are surrounded by dusty envelopes that absorb their light and re-radiate it in the infrared. When first discernible they are already well on their way to the main sequence, optically obscured by tens of magnitudes.

Are we then able to observe the newly-born stars, though indirectly? Probably not. Young as they are, and still on their way to the main sequence, the youngest observed stars must have emerged from still greater obscurity. Figure 4.9—the companion piece to figure 3.3, which illustrated the post-main-sequence adventures of a star of five solar masses—shows what theory has predicted for the development of a star of similar mass as it approaches the main sequence. When we look at the predicted temperatures we see that such a star could be optically observable only in the final stage of the journey (table 4.2). Before then it would be too cool: no stars with temperatures as low as 1000 degrees have been recorded optically. The earlier stages for a star of this mass are swiftly traversed up to

* Variable nebulous knots are so named after the astronomers George H. Herbig and Guillermo Haro, who first called attention to them.

Figure 4.6. The R Coronae Austrinae region, reproduced as a negative. Two Herbig-Haro objects are labeled, and the location of the supposed illuminating source for H-H 100 is indicated by the cross. R and T refer to R Coronae Austrinae and T Coronae Austrinae. (From Stephen E. Strom, Gary L. Grasdalen, and Karen M. Strom, *Astrophysical Journal*, 191 [1974]: 111.)

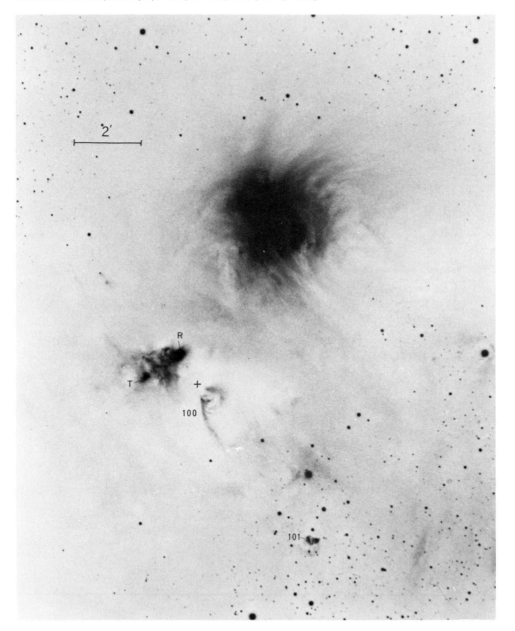

Figure 4.7. The NGC 1333 region, reproduced as a negative. Two infrared sources are indicated by crosses. The star B.D. +30°549 illuminates the bright reflection nebula in the top right corner of the photograph. The objects marked C are defects in the emulsion. (From Stephen E. Strom, Gary L. Grasdalen, and Karen M. Strom, *Astrophysical Journal*, 191 [1974]: 111.)

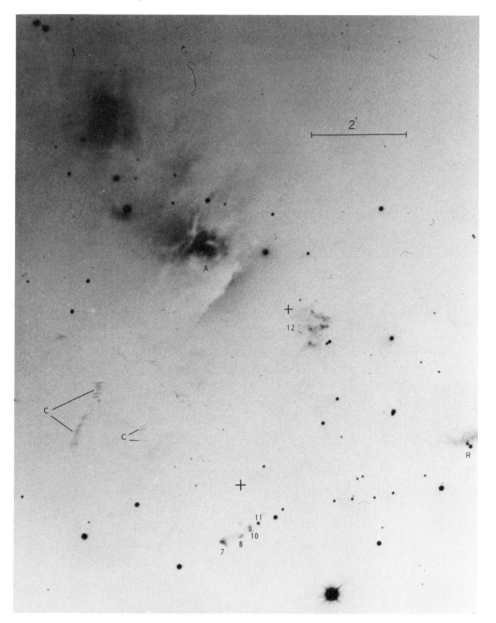

Figure 4.8.   Chart of the Chameleon cloud, showing the positions of thirty-two emission-line (pre-main-sequence) stars. The chart is reproduced as a negative. (From Karl G. Henize and E. E. Mendoza, *Astrophysical Journal*, 180 [1973]: 516.)

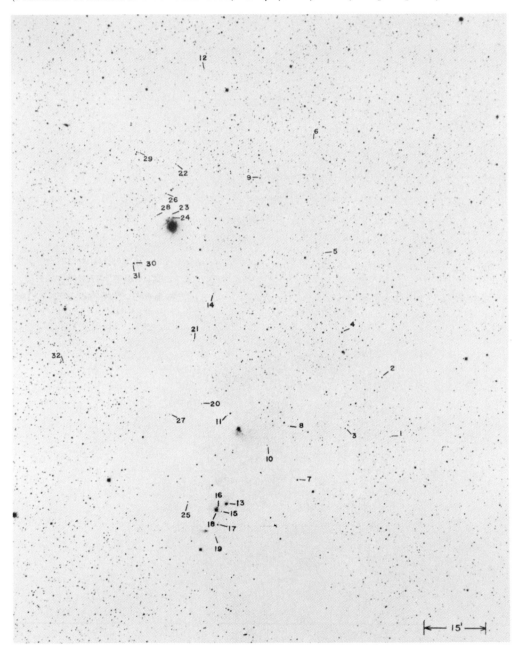

Table 4.2. Duration of pre-main-sequence phases and approximate temperature of the infalling envelope for stars of one and five solar masses.

Point on figure 4.9 at which phase ends	Duration of phase (years)		Temperature (°K)	
	1 solar mass	5 solar masses	1 solar mass	5 solar masses
1	100	100	60	30
2	1,000	1,000	90	60
3	10,000	10,000	125	300
4	100,000	100,000	300	1,000
5	1,000,000	300,000	a	3,000[b]

From Stephen E. Strom, Karen L. Strom, and Gary L. Grasdalen, *Annual Reviews of Astronomy and Astrophysics*, 13 (1975): 187.

[a] Envelope optically thin.

[b] Near main sequence.

point 4, in tens or hundreds of thousands of years, and would remain hidden from observation.

Let us look at some of the members of the Taurus association—the one first recognized by Joy—embedded in dark clouds between 300 and 400 parsecs distant. Figure 4.10 shows the relation between luminosity and effective temperature for some of its very numerous members. With few exceptions, the stars are more luminous than the sun and have about the same effective temperature; most of them are variable. In other words, they are cooler than main-sequence stars of comparable luminosity, or more luminous than main-sequence stars of comparable temperature, so they probably do not conform to the mass–luminosity relation. We believe that they are moving from right to left across the color–magnitude diagram. The marked isochrones correspond to ages of a hundred thousand and a million years; with few exceptions the stars fall between these isochrones. All of them are greatly reddened, which shows that they are deeply embedded in dust. The spectra of several of these stars are shown in figure 4.11.

Although the recorded spectra are nearly of the same general class as that of the sun, they show features unlike those of the solar photosphere. There are bright lines and a continuum of hydrogen and bright lines of ionized calcium in the optical region, as well as evidence of dust emission in the infrared. Typical spectra are shown in figure 4.11. The atmospheric motions show that the stars are expanding superficially, losing material as they do so. The higher the star in figure 4.10 (and therefore presumably the younger), the faster is it spilling off material; the older stars in the

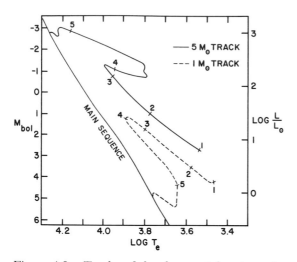

Figure 4.9. Tracks of development for stars of one solar mass and five solar masses. The coordinates are M_{bol}, the absolute bolometric magnitude (all wavelengths), $\log L/L\odot$, the logarithm of the total luminosity in terms of the sun's, and the logarithm of the effective temperature. (From Stephen E. Strom, Karen L. Strom, and Gary L. Grasdalen, *Annual Review of Astronomy and Astrophysics*, 13 [1975]: 187.)

Figure 4.10. Luminosity and effective temperature for irregular variable stars in the Taurus cloud. Luminosity is expressed as the logarithm of the ratio $L/L\odot$, the luminosity of the star in units of that of the sun; effective temperature T_e by its logarithm in °K. The zero-age main sequence (ZAMS) is indicated by a solid line, and isochrones for 100,000 and 1,000,000 years are shown by broken lines. Squares represent stars with spectra of type G or early K; triangles, stars with strong emission lines; circles, other variable stars. Those of unknown spectral type are shown by arrows. The names of known variable stars are inserted; they are in Taurus, except as noted. (Based on the diagram by A. Eric Rydgren, Stephen L. Strom, and Karen M. Strom, *Astrophysical Journal Supplement*, 30 [1976]: 307.)

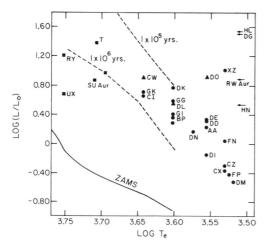

group show less conspicuous emissions in their spectra. Though their interiors must be contracting as they grow hotter at the surface and draw near to the main sequence, their envelopes are still expanding.

The stars shown in figure 4.10 are of rather low luminosity, not more than forty times as bright as the sun. Figure 4.12 shows the properties of a number of brighter pre-main-sequence stars, some of which have more than a thousand times the sun's luminosity. Their masses can be inferred from the theoretical development tracks for various multiples of the sun's mass. Some of them, such as AB Aurigae and Z Canis Majoris, are blue variable stars embedded in nebulae. The peculiarities of their spectra identify them as pre-main-sequence objects. The designation Hα, for example, indicates the presence of unusual bright lines. Although most of the known stars that are at this stage of development are faint, figure 4.12 shows that a few bright ones are observable too. Their rarity is understandable when we recall that massive stars move very rapidly through the last pre-main-sequence stages (table 4.2), the only point at which they become visible.

Figure 4.12 contains few stars near the main sequence. Are they really rare, or are such members hidden by obscuration? The question arises again when we consider the little knot of young stars in the Chameleon association (fig. 4.8). Here the two brightest stars seem to have reached the main sequence. They are not unlike Sirius, and enable us to place the association in space at a distance of nearly 400 parsecs. A glance at the mass–luminosity relation suggests that these two nests of hatching stars can contain no members of high mass. This is rather surprising, because when we study unobscured groups, the evidence suggests that the distribution of masses is always similar when stars are being formed. This, at least, is suggested by the makeup of most of the open clusters in our own galaxy. Are the Taurus and Chameleon groups exceptional, or are the more massive stars hidden deep in the dark clouds, like the infrared object in Corona Austrina?

The open cluster IC 348 suggests that this may indeed be so. Figure 4.13 shows the relation between color and brightness for this obscured and nebulous cluster. The faintest members are still evidently approaching the main sequence and include some stars with characteristic bright-line spectra. But the cluster includes two luminous main-sequence stars, and the brightest of all (o Persei, a naked-eye star) is slightly evolved. There are suggestions in the observations of another highly obscured, luminous star of high temperature.

The compact groups of irregular variables in figure 4.1 include yet other clusters embedded in bright nebulosity, such as the Lagoon Neb-

Figure 4.11. Spectra of DD, FM, CW, and HN Tauri, and HD 283447, FP, T, and GG Tauri. Each spectrum consists of several strips, which overlap slightly in wavelength; for example, the first line of the Balmer series, H α, appears to the left of the uppermost strip and to the right of the second. The red end of the spectrum is to the right. Most of the strong features are emission lines, as this spectrum is

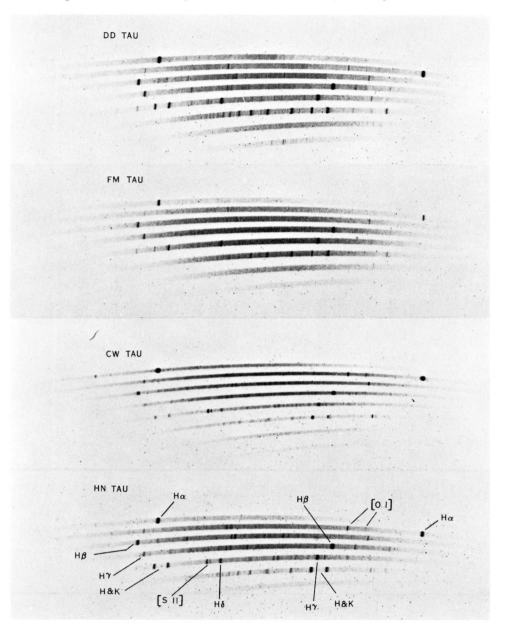

reproduced as a negative; they include the Balmer series of hydrogen, the "H and K" lines of ionized calcium, and forbidden lines of neutral oxygen and ionized sulphur. The underlying late-type absorption spectrum is also visible. (From the work of A. Eric Rydgren, Stephen L. Strom, and Karen M. Strom, *Astrophysical Journal Supplement*, 30 [1976]: 307.)

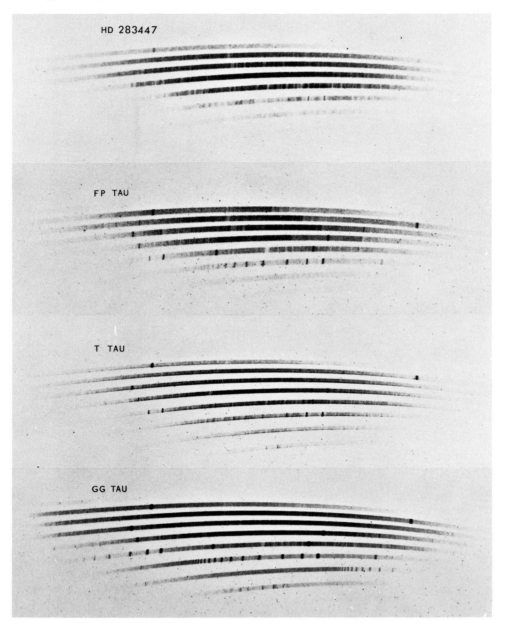

Figure 4.12.   Location of the Ae and Be pre-main-sequence stars on the luminosity–effective temperature diagram. Dots represent spectroscopic luminosity determinations; squares represent bolometric luminosities. (From Stephen E. Strom, Karen L. Strom, and Gary L. Grasdalen, *Annual Review of Astronomy and Astrophysics*, 13 [1975]: 187.)

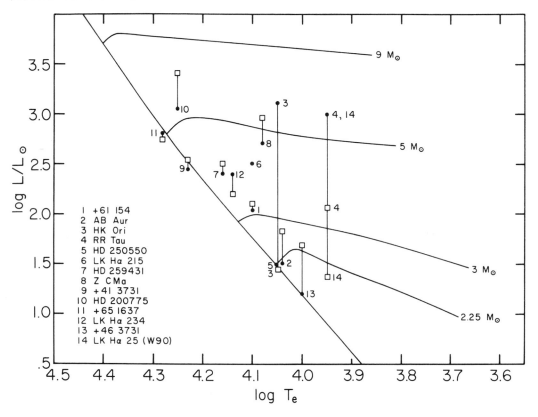

ula, Messier 20, NGC 2264, and NGC 6611. All these have well-developed main sequences, but their fainter stars are still in the pre-main-sequence stage. Even fainter stars in the still older Pleiades have not yet reached the main sequence. Indeed, a cluster as old as the Hyades is found to include a few of these contracting stars. Such clusters remind us that stars of different luminosities (and presumably of different masses) reach a given stage of development in different amounts of time.

The great complex of young stars in Perseus calls for special mention. The two compact clusters of bright stars, NGC 869 (h Persei) and NGC 884 (χ Persei), are the nuclei of the much larger Perseus II association (fig. 4.14). In NGC 869, stars fainter than absolute visual magnitude +3.5 have not yet reached the main sequence; in NGC 884, the limit in abso-

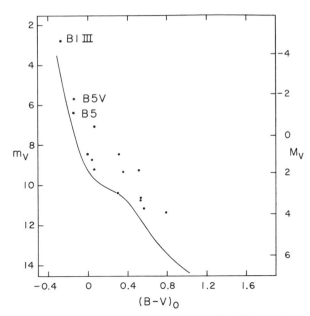

Figure 4.13. Color–magnitude array for IC 348, the very young *o* Persei cluster. Apparent visual magnitude is shown on the left, absolute visual magnitude on the right.

Figure 4.14. The double cluster in Perseus. The three color–magnitude arrays— for the two clusters and the surrounding association—carry the suggestion of slightly different ages and histories. Note that the red supergiant variables (shown by circles) are all in the association. (Based on the work of Robert L. Wildey.)

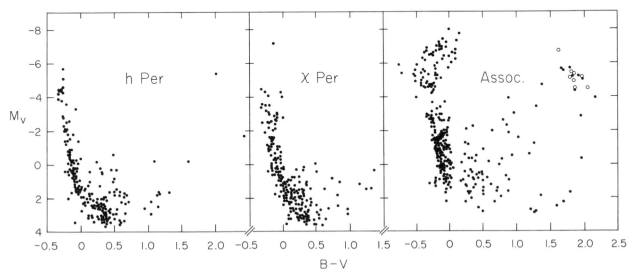

lute visual magnitude is + 2.0. The ages of the two clusters (deduced from the upper ends of their main sequences) are respectively twenty million and ten million years; brighter stars have reached the main sequence of the younger cluster. The great association that surrounds the clusters is evidently younger than either of them—or at least contains some younger stars.

A similar situation is seen in the region of Orion. A huge association of pre-main-sequence stars surrounds the bright Trapezium cluster, which lies within the great Orion Nebula in the heart of an obscuring cloud. The whole neighborhood is swarming with faint irregular variables, which must have been forming over a long time and are still being produced within the clouds.

No periodicity has ever been found for any of the T Tauri and associated stars.* Almost all other kinds of variable star show some sort of periodicity. It is possible that some of the pre-main-sequence stars may be double or multiple. In fact, so many of the stars are known to be double that it is most improbable that newly-formed objects approaching the main sequence do not share the tendency. But their observed variations of brightness furnish no evidence that can be interpreted in terms of eclipses, and their radial velocities have given no indication that they are spectroscopic binaries.

---

* The so-called BY Draconis stars, which are evidently in the pre-main-sequence stage, are periodic. Their variations are ascribed to the rotation of a star with a nonuniform surface. The known number of such stars is very small, and they are incompletely understood. Some, at least, may be spectroscopic binaries.

Creeping like Snail—The Main Sequence

CHAPTER 5

When a star reaches the main sequence, it enters on a long interval of tranquillity. The placental stage is passed; gravitational energy no longer impels the star's development. The stellar baby food—lithium, beryllium, and boron—has been consumed, and the adult digestive process begins: the star draws on the thermonuclear energy of its own substance for the first time. A brief, turbulent infancy is succeeded by a long and unruffled childhood.

At no other point in its "strange eventful history" does a star pursue its course so steadily or for so long. Single main-sequence stars do not vary in brightness, color, velocity, or spectrum—at least in their gross properties. When main-sequence stars display such vagaries, there is always evidence of the combined properties of two bodies orbiting around one another, perhaps performing periodic eclipses or showing evidence of bodily distortion. The main-sequence star has reached the first turning point on the snipelike flight that will eventually carry it back and forth across the array of stellar properties. Such a turning point always coincides with a change in diet. The rate of change during the present stage is so slow that the star appears to hang suspended before it is on its way again.

That stars spend the larger part of their lives in the main-sequence stage is emphasized by the fact that main-sequence stars are extremely numerous. Other things (such as mass) being equal, the slower a star passes through a particular stage, the more stars will be found in that stage at any one time.

The actual numbers of stars of different kinds must be deduced from a census of those known, when allowance is made for the fact that lumi-

Figure 5.1. Observed numbers of stars of different spectral classes and absolute magnitudes in volume 1 of the *Michigan Spectral Catalogue*, which includes stars south of declination − 53° (normal stars with good quality spectra). Because of the wide range in luminosity, this diagram does not show the actual numbers of stars of each kind. The catalogue has a roughly uniform limit of apparent magnitude, so we must take into account the volume of space that is included for each luminosity in order to deduce the true numbers. The result of such a calculation (the Hess diagram) is shown in figure 5.2 (which is not, however, based on the same sample of stars). (From an unpublished paper by Nancy Houk and Richard Fesen.)

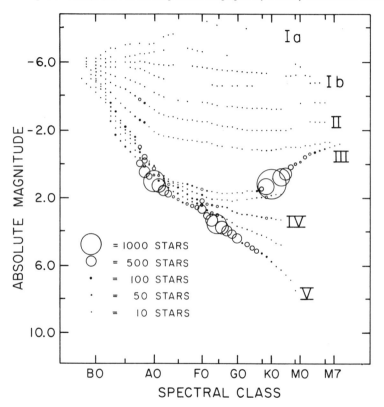

nous stars can be observed at great distances, whereas those of the lowest luminosity are only detectable in our immediate vicinity.

Figure 5.1 shows the numbers of stars actually observed, down to a fairly uniform limit of apparent magnitude, in the southern hemisphere. These data are taken from the Michigan Atlas of stellar spectra, which subdivides by spectral class and luminosity class. Concentration of stars within the main sequence (luminosity class V), is evident, even though no allowance has been made for the fact that the more luminous ones have been surveyed to the greatest distance.

Figure 5.2.   Schematic Hess diagram for stars in our neighborhood. Numbers of stars per cubic parsec are shown by contours which refer to 20, 200, 2000, and 20,000 stars. The main sequence runs along the ridge. Probably the peak is reached at about the bottom right corner of the diagram, with about 40,000 stars per cubic parsec. Statistics for fainter stars do not permit us to say how fast the slope falls off from there. The contours for white dwarfs are not shown; these stars populate a moderately high ridge, roughly parallel to the main ridge and separated from it by a deep valley.

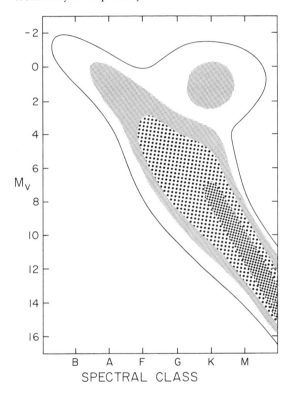

When the numbers of observed stars are corrected by considering the volume of space surveyed at each luminosity, we obtain the relative numbers per unit volume of space for each spectral class and each luminosity class. A schematic representation of the result, known as the "Hess diagram," is shown in figure 5.2, which refers to our immediate neighborhood, where the survey is reasonably complete. The contour diagram shows a conspicuous narrow ridge that runs along the main sequence, where stars of a given luminosity preponderate enormously. The ridge rises to a high point for low-luminosity stars (at least a hundred thousand times fainter than the sun), in harmony with the fact that low-luminosity stars are progressively more common than those of higher luminosity, at least down to this limit. If these main-sequence stars conform to the mass – luminosity relation, this must mean that stars of lower mass are progressively more common than those of higher mass.

The preponderance, at any level of luminosity, of main-sequence stars is a reflection of the comparatively long duration of the main-sequence phase of stellar development. Indeed, the depressions and lesser ridges of the contour diagram indicate speeds of development in the corresponding phases. Depressions mark epochs of swift development; ridges and plateaus occur where the process is slower.

The theory of stellar development springs from a knowledge of the physics of the release of thermonuclear energy under the conditions in the interiors of stars that differ in mass (fig. 3.4), and the correspondence between theory and observation is a measure of the validity of the former. The relative numbers of stars found at different points on a track for a given mass are significant, rather than the relative numbers of stars of a given luminosity. The comparison vindicates theory: a combination of the observed luminosity function with the theoretical tracks would closely reproduce the Hess diagram.

Our own sun is, of course, the best known of the main-sequence stars. It serves as a reminder of what such a star can do. It is different from other nearby and well-studied stars in that it has no known companion, while most of the others are either double or multiple. A very distant solar companion might have gone undetected, but it would have to be intrinsically faint and of very small mass. Such stars are legion. But even so, it would have to be a very distant companion indeed, or it would scarcely have escaped detection.

The possession of a planetary system may not be exceptional either—in fact, it can be regarded as a sort of multiplicity. Planets, which give no light of their own in the optical region, are hard to detect, and to get evidence of a low-mass companion calls for enormous precision. The presence of a planet as massive as Jupiter may influence the detailed behavior of the sun, but such effects on more distant stars are still beyond our observational abilities.

In the absence of compelling evidence to the contrary, then, we may regard the sun as a typical main-sequence star. It does not in fact lie precisely on the zero-age main sequence, which marks the turning point where a star begins to consume a thermonuclear diet. It has not moved far from that point, however, and is only very slightly brighter, very slightly cooler, than it was at that stage, about four and one-half thousand million years ago.

When the brightness, color, and spectrum of a star are known, they define size and surface temperature. From these data emerges a sort of thumbnail sketch of a main-sequence star: a gaseous sphere with a uniform surface that radiates light at a certain temperature and is overlaid by an absorbing atmosphere of a certain composition whose physical state and spectroscopic properties are governed by temperature and density. But a glance at the sun, even in the optical domain, reminds us that this sketch is a gross oversimplification. The sun's disk is anything but uniform. It is changing constantly and swiftly, and no one temperature can be assigned to its mottled surface. The notion of a uniform glowing sphere overlaid by an absorbing atmosphere is rudely dispelled by the phenomena of chromosphere and corona, sunspots, faculi and spicules, and "solar wind" that is continually blowing off the surface. Powerful local magnetic fields come and go with the sunspots. All this was known even from study in optical wavelengths. The picture is carried to an extreme when the far ultraviolet radiation is surveyed from a satellite outside the earth's concealing atmosphere. Little is left of the tranquil picture of a quietly radiating gaseous globe. The surfaces of all main-sequence stars may undergo comparable activity.

Changes like those observed on the face of the sun would be hard to study in the integrated light of a distant star, although they are not beyond detection. The stratification of the sun's surface, on the other hand, can be actually observed. There are: a source of continuous radiation that defines the edge of the solar disk, a layer containing absorbing atoms that lies above the disk, a hotter *chromosphere* at higher levels, and a much hotter *corona* that extends for several solar diameters. If it were not for the extraordinary chance that our moon's apparent diameter is almost exactly the same as the sun's, these details would probably not have forced themselves on our attention, for the knowledge of chromosphere and corona came initially from observations made during total solar eclipses. The higher levels of the solar envelope leap out with dramatic suddenness as the bright disk of the sun disappears behind the moon, an unforgettable revelation.

In the light of this knowledge, the stratification of the sun's surface can be analyzed and formulated in terms of a "model atmosphere," which is built up theoretically layer by layer and finally reproduces in detail the complexity and diversity of the observed spectrum. Chromosphere and corona are fitted into the picture, which has been filled out by observations of the sun's variegated surface in the far ultraviolet and even the X-ray region of the spectrum.

The model atmosphere approach can be applied to other stars, too; in conjunction with satellite observations of the ultraviolet region, it has been successful in reproducing their spectra, replacing the crude photospheric picture of an earlier day. Analysis of the spectra of stars with temperatures like that of the sun reveals the presence of chromospheres. For example, it has long been known that the centers of the strong absorption lines of ionized calcium show a delicate detail: emission lines at their centers—evidence of the presence of chromospheres. These central emissions are found not only for main-sequence stars but for larger, more developed stars of similar temperature, where they are more conspicuous. Thus, they can be used to detect such stars and even to determine their intrinsic brightness. For a few stars (though not yet for main-sequence stars) satellite observations reveal evidences of a corona in the spectra.

We can hardly doubt that other main-sequence stars can possess the analogues of sunspots; some stars not far from the main sequence do indeed display enormous magnetic fields, and evidence of spottedness can be obtained for many of them. However, with rare exceptions stars other than the sun must be studied in integrated light, and therefore surface detail may elude us. That such detail is in fact present for stars that are no longer on the main sequence, such as the α Canum Venaticorum stars discussed in chapter 6, is well established by spectroscopic and interferometric studies.

Although the envelopes of main-sequence stars must be stratified and their surfaces probably variegated, they radiate steadily and undergo no sudden or drastic changes. Their pre-main-sequence flarings and flashings are over. They do not pulsate as the Cepheid variables do. They do not undergo slow, irregular, cyclic, or periodic changes, as do many low-temperature stars. They suffer no spasmodic or explosive outbursts like the novae and supernovae. They are not in the process of a bizarre spindown like the pulsars. Many of them will eventually pass through such changes in the future course of their snipe-like flights, but the greater part of their lives—the main-sequence stage—remains uneventful.

Decades of precise work have been lavished on the study of the total energy the earth receives from the sun, the so-called "solar constant." Very small changes have indeed been noted from year to year, but they have not been shown to be intrinsic in the sun: they may be results of small changes in the transparency of our own atmosphere. The long record of biological development and evolution on the earth is incompatible with any but the most minute changes in solar radiation. There is no compelling evidence that the sun has not maintained a constant rate of radiation over many hundred million years. That there have been fluctuations in the terrestrial climate is known from the evidence of the Ice Ages, but many of these are most probably related to sunspot activity rather than to steady changes in solar radiation. In any case they have been cyclic, not secular.

When it reaches the main sequence the star emerges from obscurity. It is hard to find pre-main-sequence stars that are bright enough to be familiar, but with the main sequence we have an embarrassment of riches. The open clusters and the globular clusters furnish thousands of examples. And among the naked-eye stars there are many that are bright enough and close enough to be familiar to the watcher of the constellations. Because the bluest main-sequence stars are intrinsically the brightest, they preponderate in the list; many of the redder members, though close by, are not visible to the unaided eye. The luminosities and temperatures of a number of familiar naked-eye stars are shown in figure 5.3. Spectra of typical main-sequence stars are shown in figure 5.4.

Table 5.1 gives a list of fifteen apparently single main-sequence stars that can readily be picked out in the sky. When the eye ranges over the "eternal stars" it is sobering to remember that the oldest of these is nearly a thousand times as old as the youngest.

Three of these stars (Phecda, Merak, and Megrez) are members of the Ursa Major moving group, which includes about a hundred stars with common motion in space. A member of a much younger cluster, IC 2602, is θ Carinae. Alcor lies quite close to Mizar, a multiple system, and is probably associated with it in space. Altair is interesting because it is rotating very fast (and is therefore probably flattened, not spherical). Altair is comparatively close to us, but Deneb (α Cygni), not far from it in the sky and of about the same apparent brightness, is at least a hundred times more luminous and is accordingly more distant. It is very much younger than Altair (since supergiants develop far faster than main-sequence stars); one may surmise that ten million years ago Altair must have looked much as it does today, but Deneb was not yet in sight.

An even more evocative list of apparently bright main-sequence stars furnishes specimens that are double or multiple (table 5.2.). The table illustrates the extraordinary variety of stellar unions. Five of the systems (Spica, Algol, Alphecca, Sirius, and η Cassiopeiae) are double stars; Algol may be triple. Spica and Alphecca seem to consist of nearly identical main-sequence stars, and the components of η Cassiopeiae, though not identical, are probably both on the main sequence. But Algol consists of a main-sequence star and a subgiant that is more evolved than its primary but probably less massive. And Sirius has a white dwarf as its improbable companion. These departures from orthodoxy, by members of stellar pairs that must have had common origins, emphasize that the development of a star may be profoundly affected by the presence of a compan-

Figure 5.3. Schematic color–magnitude array for bright naked-eye stars on the main sequence. Note that the main sequence actually extends downward to about absolute magnitude 20, but there are no naked-eye stars of low luminosity.

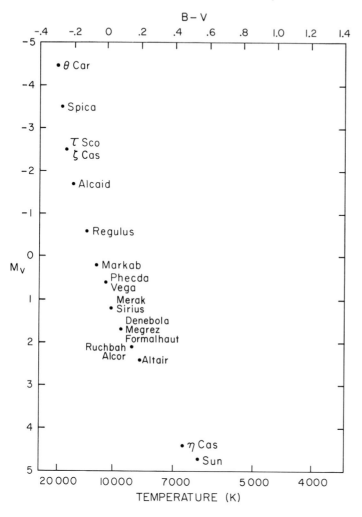

ion, even one as distant as the companion of Sirius. In view of this fact, even approximate ages have been omitted from table 5.2.

The multiple systems are still more complex and puzzling than the double stars. Mizar looks superficially like a visual double, one a main-sequence star, the other an A star with strong metallic lines. But each member of the visual double is itself a spectroscopic double star, so Mizar is at least quadruple. And Alcor may belong to the system, too.

Figure 5.4.   Spectra of typical main-sequence stars. (From William W. Morgan, Philip C. Keenan, and Edith Kellman, *Atlas of Stellar Spectra* [Chicago: University of Chicago Press, 1943]. Yerkes Observatory.)

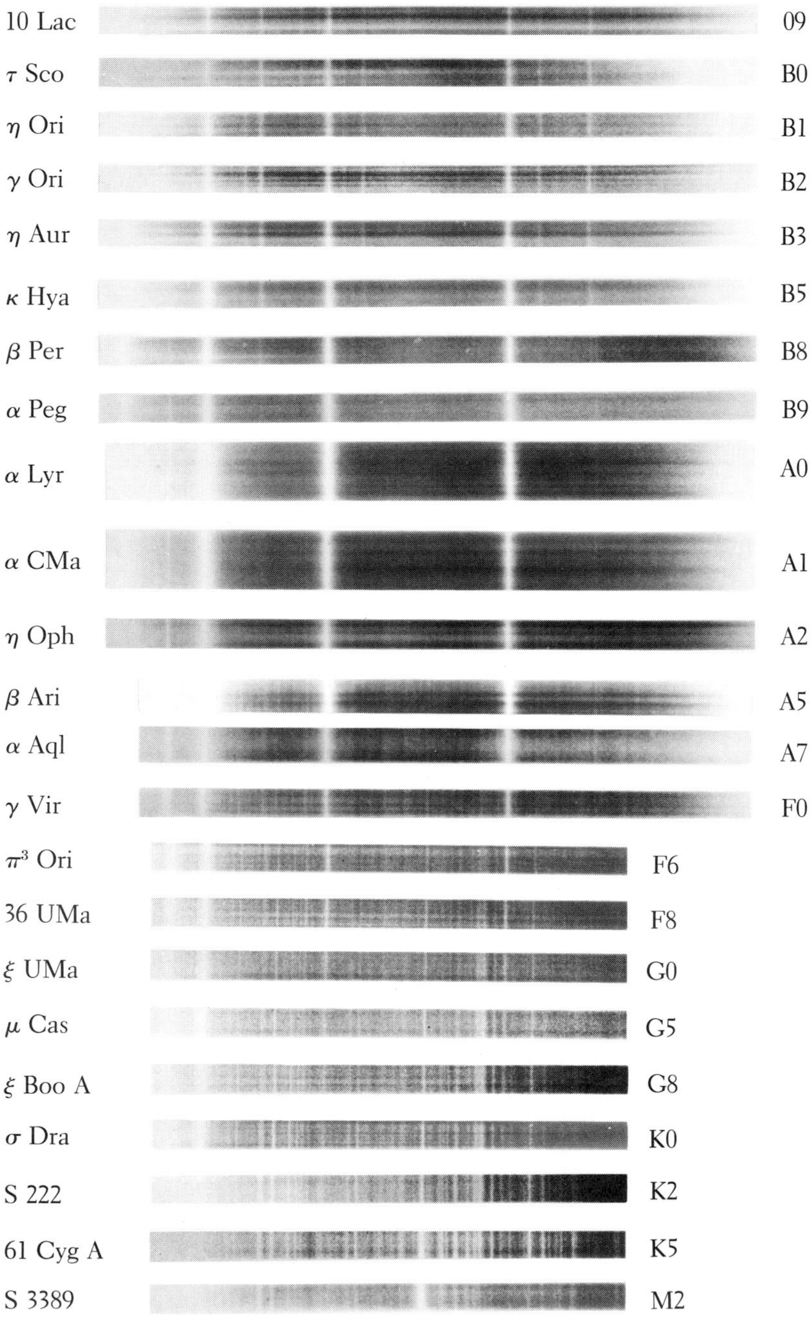

10 Lac	O9
τ Sco	B0
η Ori	B1
γ Ori	B2
η Aur	B3
κ Hya	B5
β Per	B8
α Peg	B9
α Lyr	A0
α CMa	A1
η Oph	A2
β Ari	A5
α Aql	A7
γ Vir	F0
π³ Ori	F6
36 UMa	F8
ξ UMa	G0
μ Cas	G5
ξ Boo A	G8
σ Dra	K0
S 222	K2
61 Cyg A	K5
S 3389	M2

Table 5.1. Characteristics of fifteen bright single main-sequence stars easily visible to the unaided eye.

Formal name	Familiar name	Spectrum	Apparent visual magnitude	Absolute visual magnitude[a]	Approximate age (millions of years)
θ Carinae	—	O9.5 V	2.7	−4.45	9
τ Scorpii	—	B2 V	2.82	−2.5	18
ζ Cassiopeiae	—	B2 V	3.61	−2.5	18
η Ursae Majoris	Alcaid	B3 V	1.86	−1.7	26
α Pegasi	Markab	B9 V	2.49	+0.2	165
γ Ursae Majoris	Phecda	A0 V	2.44	+0.6	230
α Lyrae	Vega	A0 V	0.04	+0.6	230
β Ursae Majoris	Merak	A1 V	2.36	+1.2	280
β Leonis	Denebola	A3 V	2.14	+1.7	420
δ Ursae Majoris	Megrez	A3 V	3.31	+1.7	420
α Piscis Austrinae	Formalhaut	A3 V	1.16	+1.7	420
δ Cassiopeiae	Ruchbah	A5 V	2.68	+2.1	660
80 Ursae Majoris	Alcor	A5 V	4.01	+2.1	660
α Aquilae	Altair	A7 V	0.77	+2.4	900
—	Sun	G2 V	−26.86	+4.7	5,000

[a] Average for the spectral class.

The sextuple system of Castor is especially noteworthy, though there is no reason to suppose that it is exceptional. The two brighter components, listed in the table, are both spectroscopic binaries, with periods of 9.2 and 2.9 days respectively. The spectrum of the brighter places its components on the main sequence; the fainter is an A star with strong metallic lines— a combination that recalls Mizar. The two brighter binary components of Castor themselves form another binary, orbiting around one another in a period of 380 years. The third visual component (Castor C, also known as YY Geminorum on account of its variable brightness) consists of a pair of main-sequence M stars so oriented that they are observed to eclipse with an orbital period of about 19.5 hours. They are a part of the Castor system but are distant from the other four stars, and the period of their orbit around the latter must be millions of years.

The nearest system to us, that of α Centauri, contains one star very similar to the sun; its visual companion is smaller, cooler, and less massive. It does not seem to lie on the main sequence but to be slightly evolved. The third member of the group, Proxima Centauri, is slightly variable. It might be surmised to be still approaching the main sequence,

Table 5.2. Characteristics of twelve bright double and multiple main-sequence stars easily visible to the unaided eye.

Formal name	Familiar name	Spectrum		Apparent visual magnitude	Absolute visual magnitude[a]	Remarks
		Main-sequence component(s)	Companion			
β Scorpii	—	B0.5 V, B2 V		2.63, 4.92	-3.8, -2.5	Multiple system
α Virginis	Spica	B1 V, B1 V		0.96	-3.5	Eclipsing star
δ Orionis	Mintaka	B2 V	O9.5 II	6.87, 2.20	-2.5	Triple
α Leonis	Regulus	B7 V	K1	1.36	-0.6	Multiple
β Persei	Algol	B8 V	G8?	2.2	-0.2	Triple, eclipsing
α Coronae Borealis	Alphecca	A0 V		2.23	+0.6	Eclipsing star
α Draconis	Thuban	A0 V		3.64	+0.6	Spectroscopic binary
α Geminorum	Castor	A1 V	Am, d M1	1.99, 2.85	+1.2	Sextuple
α Canis Majoris	Sirius	A1 V	White dwarf	-1.47, 8.5	+1.2	Visual binary
ζ Ursae Majoris	Mizar	A2 V	Am	2.40, 3.96	+1.4	Two spectroscopic binaries
η Cassiopeiae	—	G0 V	d M	3.45, 7.7	+4.4	Multiple
α Centauri	—	G2 V	d K1, d M	0.33, 1.70, 14	+4.7	Visual triple

[a] Average for main-sequence component.

but its spectrum does not seem to resemble the spectra of known pre-main-sequence stars.

Mintaka, a member of a triple system, differs from the other stars in table 5.2 in having a companion that is much brighter than itself, a so-called "bright giant," which actually gives most of the light of the system. It is over four magnitudes brighter than the main-sequence component and is itself a spectroscopic binary and perhaps an eclipsing star.

Table 5.2 was compiled from among the brightest naked-eye stars. That even this list contains a number of strange bedfellows is a reminder that the task of tracing stellar development and relationships will not always be an easy one.

Here we take leave of the main sequence, the long stellar childhood. Against the background of the star clusters, groups of similar age and history, we shall now watch the progress of stars as they leave the main sequence and move toward maturity and old age.

Leaving the Main Sequence

A star of five solar masses is about five hundred times as luminous as the sun when it reaches the main sequence, and a great deal hotter. Visually it is less luminous than this; because of its high temperature much of the light it radiates is in the ultraviolet region, beyond the visual range. Physically it would be somewhere between Alcaid and the brighter component of Regulus, with spectrum of class B5 V and a surface temperature about 16,000 degrees. It would resemble the naked-eye star ν Andromedae. We can suppose that soon after it became a visible star, it passed through a stage like that of Z Canis Majoris on the way to the main sequence, and that it was not much fainter during that stage than when it arrived at the main sequence, but was considerably cooler.

Such a massive star spends a relatively short time in childhood tranquillity, about sixty million years as compared to several thousand million years for a star like the sun. It is consuming hydrogen, first in the central core and then in a growing internal shell, leaving helium, the product of the digested hydrogen, at its core. As the hydrogen is converted into helium, releasing energy in the form of radiation, the helium core grows steadily. The star responds by becoming slightly more luminous and slightly cooler at the surface. It has passed the first bend in the developmental track, point 2 in figure 3.3.

Still burning hydrogen in a growing internal shell, a star of five solar masses now begins a bodily contraction, and for about two million years after point 2 in the track it grows slightly brighter, slightly hotter. The details of the star's inner economy are too complex to describe here: they depend critically on the degree to which the contents of the star's interior are mixed. The calculations whose results are illustrated in figure 3.3 indicate that after about two million years the star begins to expand again and passes through point 3 in its track under the impetus of the energy liberated in a growing hydrogen-burning shell. In little more than a million years it will become a red giant at point 5. The tracks by themselves are somewhat deceptive, for the speed with which the star follows them undergoes drastic changes. From the first point (the zero-age main sequence) to the second, the time interval is thirty times that from the second point to the third; and in moving from the third point to the red-giant stage the star spends only one fiftieth of its main-sequence lifetime. This helps us to understand why red giants are so much rarer than main-sequence stars of similar mass.

Figure 3.4 shows that the form of the tracks is not the same for stars of all masses. The little crotchet at the edge of the main sequence is not predicted for stars of solar mass or less.

A graphic illustration of the early stages of post-main-sequence development is shown in figures 6.1 through 6.11 (see also table 6.1). Here are the color–magnitude arrays of ten open clusters, arranged in order of the (visual) luminosities of their brightest stars.* Their ages, deduced from the luminosity at which the members begin to deviate from the main sequence, range from about fifteen million years for NGC 2362 to about one hundred fifty million for the Ursa Major group Collinder 285.

It is well to emphasize that the ages of clusters that are mentioned here and elsewhere are but rough approximations. The details of the predicted tracks and the speed with which the star develops are very sensitive to composition, and we have already noted small difference of composition between open clusters. In the overall picture we have aimed at consistency. However, different investigators have assigned somewhat different ages even to well-studied clusters. A series of clusters arranged in order of the luminosity of the stars at the top of the main sequence is probably also in order of age, although difference of composition may distort this order slightly. We should recall also that age, as here used, refers to the interval since the zero-age main sequence, not since initial formation. If we wish to count from that elusive point, the times should be increased by about 10 percent.

Each cluster in figures 6.1 through 6.11 conforms well to the main sequence in the middle range of magnitudes. But in each picture we note that the brightest stars are deviating slightly to the right of the main sequence. We may suppose that these stars have begun to build up a central core of helium and to move away from the zero-age main sequence. For

---

* Luminosity, as used here, always implies absolute brightness.

Figure 6.1.   NGC 2362. The brightest star is τ Canis Majoris. The very red variable VY Canis Majoris, a possible supergiant member, would be far outside the limits of the diagram, as indicated by the arrow. In this color–magnitude array, the apparent visual magnitude (m_v) and absolute visual magnitude (M_v) are plotted against color index $(B - V)$.   (Photograph by Harvard Observatory.)

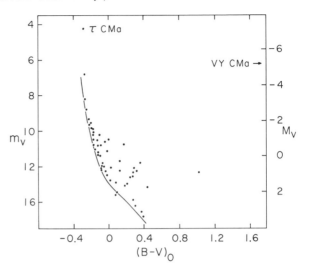

Figure 6.2.   IC 1805. The brightest stars are supergiants (luminosity class Ia) surrounded by dust shells. The whole cluster is streaked with nebulosity. The small circles represent the mean luminosity of a number of stars that are approaching the main sequence. The broken lines are isochrones corresponding to ages of one million and three million years respectively. Stars of luminosity class V are denoted by V. (Based on the work of A. F. J. Moffat; photograph by Harvard Observatory.)

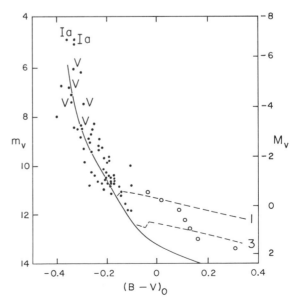

Figure 6.3.   NGC 7380. The two brightest stars are blue supergiants. The brighter of the two is the eclipsing binary DH Cephei, consisting of two O stars with about twenty times the sun's mass. Circles refer to spectroscopic binaries. (Photograph by Harvard Observatory.)

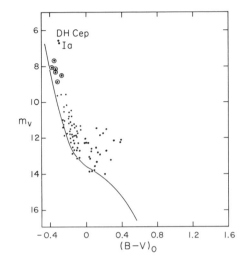

Figure 6.4.   NGC 2264. The brightest star is the irregular variable S Monocerotis. (Photograph by Harvard Observatory.)

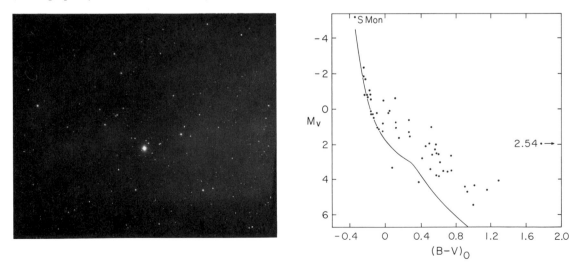

Figure 6.5.   NGC 2232. The brightest member is the naked-eye star 10 Mono-cerotis, a visual binary system. (Based on the work of J. J. Claria; photograph by Harvard Observatory.)

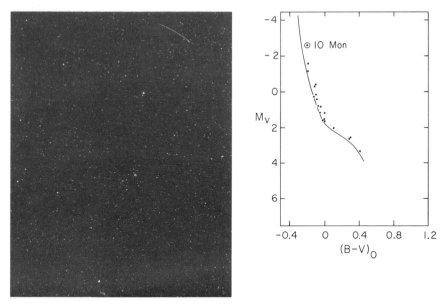

Figure 6.6.   NGC 6025. This diagram, which has been corrected for reddening, shows that most of the brighter stars have moved slightly away from the main sequence (compare the Pleiades, fig. 6.7). The brightest star shows emission lines. (Based on the work of A. Feinstein; photograph by Harvard Observatory.)

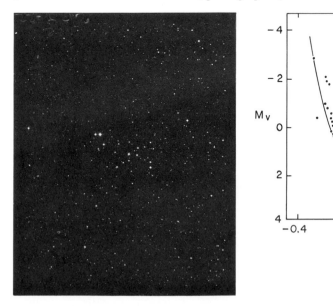

Figure 6.7.   Color–magnitude array for the Pleiades. The brightest stars are veering away from the main sequence. Circles denote spectroscopic binaries; the cross refers to Pleione.

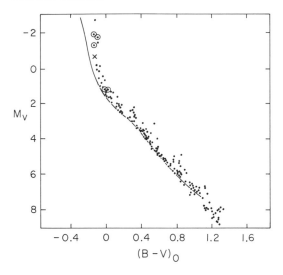

Figure 6.8.   The Pleiades. Top left: a photograph of the brightest stars. Top right: a longer exposure showing the reflection nebulae in which the stars are enmeshed. Reading from left to right in the upper left picture: Atlas, Pleione (very faint), Alcyone, Merope (deep in nebulosity), Maia, Electra, Taygeta. Asterope is too faint to show in the left-hand picture; it can be seen near the top center of the right-hand photograph. Celaeno is outside the limits of the photograph. Below: a photograph of the spectra of the six brightest stars, which show primarily the Balmer lines of hydrogen. (Photographs by Yerkes Observatory.)

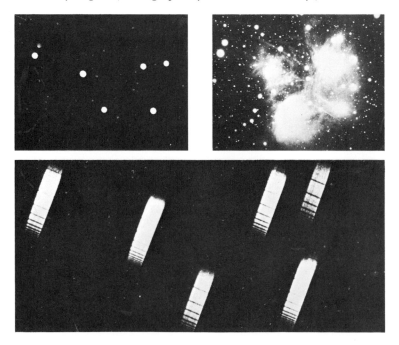

Figure 6.9.   NGC 6405. The brightest star is the semiregular supergiant variable
BM Scorpii. (Photograph by Harvard Observatory.)

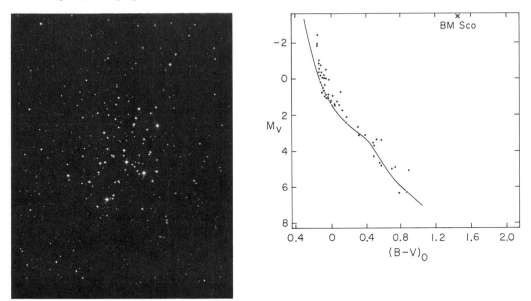

Figure 6.10.   NGC 2168. The red giant is probably an evolved member of the
cluster. (Photograph by Harvard Observatory.)

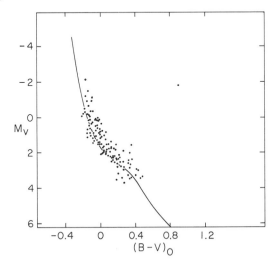

Figure 6.11. The Ursa Major group. The circled star is ε Ursae Majoris, an α Canum Venaticorum variable. This is a moving cluster, and no picture can be shown. The distances and luminosities of the stars have been individually determined.

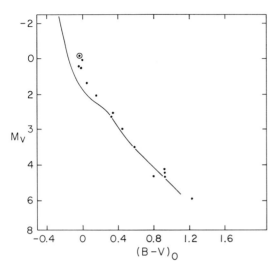

Table 6.1. Positions of fifteen open clusters that include stars in the early stages of post-main-sequence development. Clusters bright enough to be readily seen with the naked eye or binoculars are marked with asterisks.

Name	Position			
	h	m	°	′
IC 1805	2	25.2	+61	0
*Pleiades	3	41	+23	48
*NGC 1502	3	58.7	+62	3
*NGC 2168 (M35)	6	2.7	+24	21
NGC 2232	6	21.3	− 4	40
NGC 2264	6	35.5	+ 9	59
Cr 121	6	50.0	−24	30
*NGC 2362 (τ CMa)	7	14.6	−24	26
Hogg 15	12	40.6	−62	50
NGC 6025	15	56.3	−60	14
*NGC 6231	16	47.0	−41	38
NGC 6383	17	28.2	−32	30
*NGC 6405 (M6)	17	33.5	−32	9
NGC 6871	20	2.1	+35	30
NGC 7380	22	43.0	+57	34

NGC 2362 the swing away from the main sequence is only just percepti-
ble (fig. 6.1). By the time we reach the Pleiades it is well marked (fig. 6.7),
and in the sparsely populated Ursa Major cluster it has gone far indeed
(fig. 6.11). In looking at these diagrams we must keep in mind the difficulty of determining luminosity and color and of assigning individual
stars to groups that was already encountered with the best-known of all
clusters, the Hyades.

The extreme youth of NGC 2362, IC 1805, and NGC 2264 is attested
by the large number of their faintest stars that have not reached the main
sequence. NGC 2264 (fig. 6.4) is a well-known nest of irregular variables
like the pre-main-sequence stars described in chapter 4. IC 1805 is permeated with filaments of bright hydrogen nebulosity, and its most luminous and massive members seem to be surrounded by dust shells (fig.
6.2). One of the brightest has a spectrum that points to a very high mass
—over fifty suns. Its faintest members are evidently still on their way to
the main sequence. These are shown in the diagram by circles which represent the luminosity of a number of stars. They fall between theoretical
isochrones that correspond to intervals of one million and three million
years since first formation. The age of this cluster, deduced from the
point of deviation of its brightest stars from the main sequence, is between one and two million years, so there is an indication that star formation is still taking place in IC 1805. Most of the members whose spectra
are known show marks of main-sequence membership, but the three
brightest are supergiants that have begun to move to the right. This cluster contains a compact multiple system whose age must be less than a million years; it is obscured by more than two magnitudes, and the stars are
correspondingly reddened.

Double and multiple stars play a conspicuous part in these clusters.
The brightest member of NGC 2362 is the fourth-magnitude star τ Canis
Majoris, a spectroscopic binary with an orbital period of 158 days (fig.
6.1). Another interesting star that may be a member of the cluster is the
very red irregular variable VY Canis Majoris, which is embedded in dust
and surrounded by an expanding nebula. At its brightest it is visually
about as bright as τ Canis Majoris. If it is also as massive, it could have
reached its present stage as a luminous red star in something like a million
years. Such very red bright stars are rare in young clusters, but this rarity
is to be expected if indeed they develop as fast as theory suggests. Figure
6.1 does not show VY Canis Majoris in the color–magnitude array; it
would be far beyond the right-hand edge, and because of the enmeshing
dust cloud its true color is hard to determine.

The brightest member of NGC 7380 is the eclipsing variable DH Cephei, a pair of nearly identical O5 stars with an orbital period of 2.11 days

and masses of about twenty-three and nineteen suns (fig. 6.3). The brightest member of NGC 2264 is the double star S Monocerotis, whose brighter member is an irregular variable, possibly a late-born star still approaching the main sequence. The brightest star in NGC 2232 (10 Monocerotis) is a visual binary, and one member of the pair is a spectroscopic binary (fig. 6.5). In NGC 6025 the brightest star is a hot star with emission lines in its spectrum, the sign of a glowing envelope perhaps confined to a rotating equatorial ring (fig. 6.6).

A whole book could be written about the Pleiades, whose brightest stars are a lovely group to view, even with the unaided eye. All the naked-eye members have swung away from the main sequence (fig. 6.7). They are enmeshed in nebulosity, which (as its spectrum shows) shines by their reflected light (fig. 6.8). Three of the brightest members and several fainter ones are known to be spectroscopic binaries. All are rotating rapidly. Merope, the one most deeply embedded in obvious nebulosity, has a striking bright-line spectrum. Maia may be slightly variable. The most strikingly variable is the rapidly rotating Pleione (BU Tauri), whose spectrum shows that it goes through episodes in which it ejects shells of material, a not uncommon trick of bright, hot stars. The second magnitude γ Cassiopeiae has had spectacular episodes of this sort, and it too is rotating rapidly and is slightly variable in brightness.

Further down the main sequence of the Pleiades we note something already seen in the Hyades: a number of stars lie on a line above and roughly parallel to the backbone. These, like the similar stars in the Hyades, are undoubtedly double, and a few spectroscopic binaries are actually known in this range of brightness.

At the lower end of the main sequence the stars of the Pleiades appear to conform less closely. This is not only (though it may be partly) due to observational uncertainty. At about the thirteenth apparent magnitude, absolute visual magnitude near 7, we have reached the point at which stars are still approaching the main sequence. A very large number of typical irregular variables have been found at this magnitude and below it, stars such as FL Tauri and V 703 Tauri, both with spectra of class K. There is little doubt that most of them are actually associated with the cluster and still in their infancy.

In the Pleiades we meet for the first time the regularly periodic intrinsically variable stars. Four δ Scuti stars (so named for the brightest member of the group) are found in the cluster. Their periods of pulsation or vibration are near to an hour. This behavior is still something of a mystery (see chapter 8).

The clusters illustrated in figures 6.1 through 6.11 are only specimens, each with its own individuality. Figures 6.12 and 6.13 show two more very

Figure 6.12. NGC 1502. Top left: color–magnitude array. One of the two brightest stars is the eclipsing variable SZ Cameloparadalis. The diagram above and to the right shows its observed light variations. The two halves of the curve are quite similar and have been reflected about the primary minimum at phase 0; dots denote the first half of the curve, circles the second. The two stars are so close together that they distort each other; in order to deduce their properties it is necessary to rectify the light curve by removing the effects of this distortion. The rectified curve is shown below the observed curve. At bottom are two views of the system as seen from our position: midway between eclipses and at primary minimum (the deeper of the two). The size of the sun is shown for comparison. (From the work of Adriaan Wesselink.)

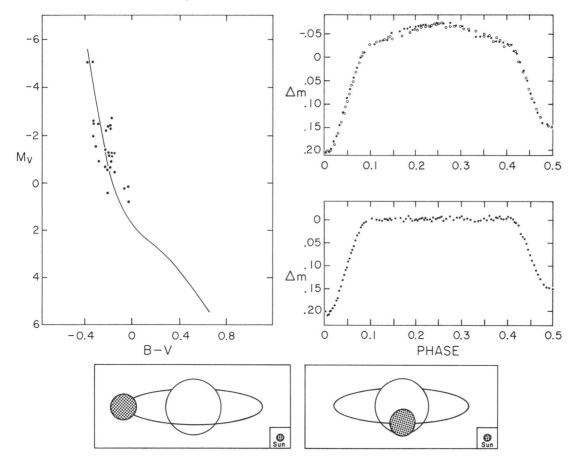

young clusters, NGC 1502 and NGC 6383, and the binary stars associated with them. The eclipsing star SZ Camelopardalis, the brightest member of NGC 1502, is a pair of hot B stars with an orbital period 2.698 days. The brightest member of NGC 6383 is a spectroscopic binary with a period of 3.66 days; it consists of a pair of similar O stars and is possibly variable.

Figure 6.13. Color–magnitude array for NGC 6383 and the light curve of the member star V 701 Scorpii, marked in the array with a circle. The upper light curve is recorded in yellow light, the lower in blue light; above the two curves is a record of the color. The constancy of color shows that the two components of this eclipsing binary are of the same temperature; the form of the curves shows that they are similar in size and so close together as to distort each other. The orbital period is about 18 hours. (The light curve is by Kam-Ching Leung, *Astronomy and Astrophysics Supplement*, 13 [1974]: 315.)

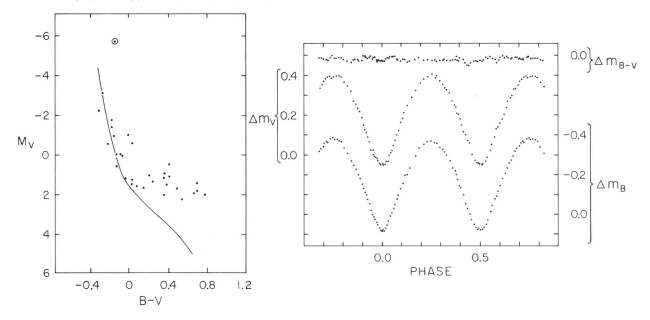

The masses are probably at least twenty times the sun's. The second brightest star in the cluster is also double, the eclipsing star V 701 Scorpii, with two similar B5 members and an orbital period of about 18 hours. The light variations show that they are almost identical and that they are revolving almost in contact.

In each of the young clusters that have been examined, the brightest, most massive stars are binary or multiple and may be considered to form the nuclei of the clusters. Several of the young clusters are found, in turn, to form the nuclei of more extended young associations. A striking example is furnished by the twin clusters in Perseus (fig. 6.14), which form the nuclei of a much larger loose association of young stars. The associations are not gravitationally bound and will tend to dissipate with advancing age, so that we shall find that older clusters are more isolated. Many of the stars that formed the original associations must be sought in the surrounding star fields. A transitional situation has been described for

Figure 6.14.   The double cluster in Perseus, associated with a number of β Canis Majoris stars. (Photograph by Harvard Observatory.)

the relatively old Hyades cluster, surrounded by the gradually dispersing Hyades group, which has about ten times the cluster's diameter. We shall find that the very oldest open clusters tend to be inconspicuous, both because all but the most populous have lost a large proportion of their original members in the va-et-vient of their travels within the Galaxy and because their brightest and most massive stars have run their course and passed from the scene. Their main sequences are represented only by stars of lower mass and luminosity.

Before leaving the young clusters, we may consider the light they throw on the enigmatic group that comprises the rare and spectacular Wolf-Rayet stars (named for the German and French astronomers who were among the first to study them). The spectra of most stars show primarily (though not exclusively) absorption lines, but the Wolf-Rayet stars have primarily (though not exclusively) emission spectra, which suggest that they are surrounded by brilliantly emitting envelopes that seem to be in violent motion, perhaps also in expansion. Many are spectroscopic binaries with luminous, high-temperature companions. A great breakthrough in understanding the Wolf-Rayet stars came when several were found to be eclipsing binaries, for it opened the way to a knowledge of their masses and dimensions. But we are still far from understanding them. Their atmospheric structure is extraordinarily complex, and some of the binaries have shown remarkable changes from cycle to cycle.

All the Wolf-Rayet stars in our galaxy are near the central plane, and many (but not all) form loose groups, notably in Cygnus and in Carina. They must, therefore, be placed with the young stars, though not perhaps with the very youngest. The question arises whether they are going or coming with respect to the main sequence: are they pre- or post-main-sequence stars? Their presence in a few open clusters can throw light on this question.

Figure 6.15 shows the makeup of four open clusters that contain Wolf-Rayet stars. The two Wolf-Rayet stars in NGC 6231 (a compact cluster that is one of the nuclei of the great association Scorpius II) are spectroscopic binaries. All four clusters are manifestly quite young. Their pre-main-sequence stars, if any, are many magnitudes fainter than the Wolf-Rayet stars, which are all at or near the top of the main sequence. (The spectra of these stars are so peculiar that their colors must deviate from the orderly pattern that represents ordinary stars, and luminosity rather than color is significant here.) Similar observations have been made of the Wolf-Rayet stars in several clusters of the Large Magellanic Cloud, a stellar system richer than our own in such objects.

My opinion is that all Wolf-Rayet stars are probably binaries and owe their spectacular properties to interplay between the hot components. But this is not the whole story, for some Wolf-Rayet stars are members of well-separated binaries. And why, for example, is the equally hot DH Cephei not a Wolf-Rayet star? Whatever the source of their peculiarities, their position in the color–magnitude diagrams of open clusters, near the first turning point and at the tip of the main sequence, suggests that these stars are leaving it, or about to leave it.

In following the adventures of a star of five solar masses, we noted a little twist or crotchet in the development track at the upper end of the main sequence. Is this change in the course of stellar development marked by any change in stellar behavior? An answer to this question may

Figure 6.15.   Color–magnitude arrays of four clusters that contain Wolf-Rayet stars: NGC 6231, 6871, Hogg 15, and Collinder 121. The Wolf-Rayet stars are denoted by circles. Photographs of NGC 6231 and Collinder 121 are shown below. (Photographs by Harvard Observatory.)

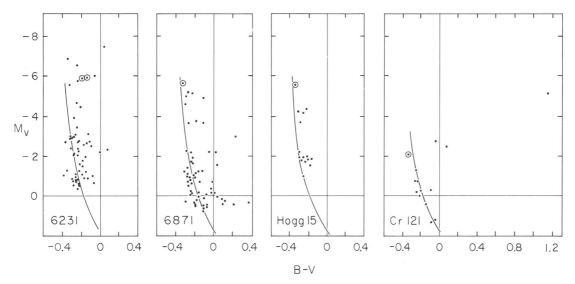

be found in the two rich young clusters NGC 869 (h Persei) and NGC 884 (χ Persei). It appears that the upper edges of their main sequences are populated by representatives of the group of variable stars known (from one of the brightest members) as β Canis Majoris stars. All are early B stars, and all belong to luminosity classes III and IV, which means that they are slightly evolved away from the main sequence. Their variations of brightness are small and rapid, with periods from two hours to about half a day, and are also very complex. Many show two nearly equal periods that alternately reinforce and cancel the changes of brightness. There are concurrent variations of spectrum and radial velocity. Some of them, though probably not all, are spectroscopic binaries.

An attractive suggestion is that the two contending periods are the natural periods of vibration of the equatorial and polar diameters of stars that are appreciably flattened by rapid rotation. As we scan the main sequence we notice that the rapidly rotating representatives are all young, presumably massive stars; older stars rotate progressively less and less rapidly. The sun, for example, rotates comparatively slowly.

Although the nature of the complex vibrations is still a matter of debate, and what activates them is not clearly understood, it is evident that the β Canis Majoris stars lie in the region of the color–magnitude diagram where the development tracks are undergoing the kink that is theoretically predicted, as illustrated in figure 6.16.

Because their variations are small and rapid, these stars are not readily found (fig. 6.17). The majority of those now known are bright stars, which can most easily be studied in detail. A list of such stars that can be readily picked out with the unaided eye is given in table 6.2. They are confined between spectral classes B0.5 and B3, and almost without exception are of luminosity classes IV, III, or II, so they are not on the main sequence. It is quite possible that all stars within these limits of spectrum and luminosity will ultimately prove to be β Canis Majoris stars. Indeed, one of the hot stars that has been most intensively studied, because it was thought to be a typical and stable representative of its class (B2 IV), γ Pegasi, has proved to be a β Canis Majoris star. So has the first magnitude star β Crucis, one of the brightest in the sky.

The β Canis Majoris stars now known are confined to the hottest spectral classes. Is there any evidence of unusual behavior among cooler stars that is related to the kink? Reference to figure 3.4 shows that the kink can be traced in the theoretical curves for stars with masses down to about 1.25 times that of the sun, which corresponds on the main sequence to about spectral class F0. Evidence must be sought for stars with spectra between B and F.

Figure 6.16.   Relation of the β Canis Majoris stars to the main sequence (heavy line). The domain of the variables is indicated by light lines, and the twist in the development track is shown.

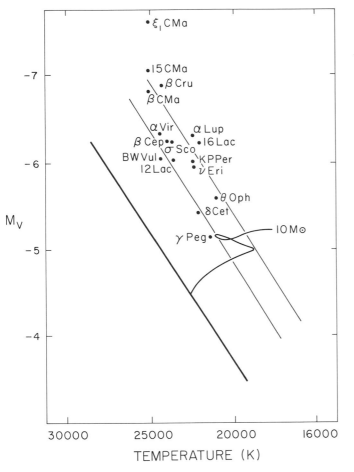

The Ursa Major cluster includes indeed a representative of another group of variable stars of small range, the α Canum Venaticorum stars, named for its brightest member, Cor Caroli (fig. 6.11). Their periods are longer than those of the β Canis Majoris stars, between one and twenty-five days. Here again most of the known specimens are fairly bright. A list of those easily visible to the unaided eye is given in table 6.3. Their spectra range from B5 to F0, just the range we were seeking.

The spectra of most of these stars are classed as *p*, meaning "peculiar"; they are marked by the unusual intensity of many atomic absorption lines, notably silicon, strontium, and the rare earths, and they vary in a

Figure 6.17. Variations of the β Canis Majoris star BW Vulpeculae on four nights. The magnitudes are expressed relative to the magnitude of a comparison star of constant brightness. (From C. Roger Lynds, *Publications of the Astronomical Society of the Pacific*, 66 [1954]: 199.)

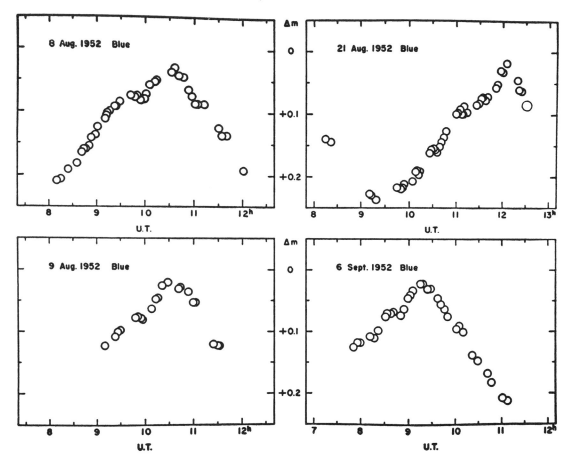

peculiar manner. In step with the variations of brightness and spectrum, these stars display evidence of periodically fluctuating magnetic fields, often of great intensity. There is little doubt that the periods are those of rotating stars whose magnetic poles do not coincide with the poles of rotation and that the changes observed are presentation effects: we view different parts of the star's surface as it rotates. In fact, it is possible, by an analysis of the magnetic structure of the spectrum lines, to obtain an idea of the surface variegations of these stars. Although the α Canum Venaticorum stars seem to lie in the domain of the color–magnitude diagram

Table 6.2. Characteristics of twenty-one β Canis Majoris stars easily visible to the unaided eye.

Name	Max.	Min.	Spectrum	Period (days)	DM[a]		Position (1900)				
							h	m	s	°	'
γ Peg (88 Peg)	2.80	2.87	B2 IV	0.157495	+14°	14	0	8	5	+14	37.6
δ Cet (82 Cet)	3.80	3.86	B2 IV	0.1611366	− 0	406	2	34	21	− 0	6.2
ν Eri (48 Eri)	3.40	3.60	B2 III	0.17790414	− 3	834	4	31	19	− 3	33.4
β CMa (2 CMa)	1.93	2.00	B1 II-III	0.25002246	−17	1467	6	18	18	−17	54.4
ξ¹ CMa (4 CMa)	4.33	4.36	B1 III	0.2095755	−23	3991	6	27	41	−23	20.8
EY CMa (15 CMa)	4.60	4.63	B1 III	0.184557	−20	1616	6	49	13	−20	6.0
FN CMa	5.36	5.39	B0.5 IV	0.12377	−11	1790	7	1	59	−11	8.4
o Vel	3.56	3.67	B3 III-IV	0.131977	−52	1583	8	37	26	−52	34.0
β Cru	1.23	1.31	B0.5 III	0.2365072	−59	4451	12	41	52	−59	8.5
ε Cen	2.30	2.31	B1 III	0.169608	−52	6655	13	33	33	−52	57.5
τ¹ Lup	4.36	4.43	B2 IV	0.177365	−44	9322	14	19	43	−44	46.2
α Lup	2.28	2.31	B1.5 III	0.259864	−46	9501	14	35	17	−46	57.6
δ Lup	3.21	3.24	B2 IV	0.16547	−40	9538	15	14	48	−40	17.1
σ Sco (20 Sco)	2.94	3.06	B1 III	0.2468406	−25	11485	16	15	6	−25	21.2
θ Oph (42 Oph)	3.25	3.29	B2 IV	0.140531	−24	13292	17	15	52	−24	54.1
λ Sco (35 Sco)	1.59	1.65	B1 V	0.2137015	−37	11673	17	26	49	−37	1.9
κ Sco	2.39	2.42	B2 IV	0.19987	−38	12137	17	35	34	−38	58.7
V 2052 Oph	5.81	5.84	B2 IV-V	0.1398903	+ 0	3813	17	51	13	+ 0	41.1
β Cep (8 Cep)	3.30	3.35	B1 III	0.1904844	+69	1173	21	27	22	+70	7.3
DD Lac (12 Lac)	4.90	5.10	B1.5 III	0.19308858	+39	4912	22	37	0	+39	42.2
EN Lac (16 Lac)	5.30	5.41	B2 IV	0.169165	+40	4949	22	51	50	+41	4.2

[a] In this and later tables, DM refers to the Durchmusterung number: the *Bonn Durchmusterung* for stars north of −23°, the *Cordoba Durchmusterung* for stars south of −23° and down to −52°, and the *Cape Photographic Durchmusterung* for stars from −52° to the South Pole. Stars are numbered serially within each zone of declination. On account of precession, the declinations of some stars have moved into the neighboring zone of declination since the catalogues were made. The *Bonn Durchmusterung* gives positions for 1855, the two southern catalogues, for 1875.

where the crotchet occurs, it is not clear why the observed remarkable variations should be associated with this phase of the stellar lifetime.

Of the fifty-seven known α Canum Venaticorum stars, 9 percent are in open clusters and 14 percent are known to be binary. The sample is too small to permit definite conclusions. Two are in Coma Berenices, one in the Ursa Major cluster, one in the Perseus moving cluster, and one, possibly, in the Pleiades. The most that can be said is that their distribution is suggestive, in that it recalls that of the β Canis Majoris stars.

The stage of development represented by clusters whose brightest

Table 6.3.   Characteristics of thirty-six α Canum Venaticorum stars easily visible to the unaided eye.

Name	Max.	Min.	Spectrum	Period (days)	DM		Position (1900)				
							h	m	s	°	'
α And (21 And)	2.02	2.06	B8 IVp	0.9636	+28°	4	0	3	13	+28	32.2
γ Ari (5 Ari)	3.88	3.90	A0 p	2.607	+18	243	1	48	3	+18	48.1
ι Cas	4.50	4.53	A5 p	1.74058	+66	213	2	20	49	+66	57.2
LT (Per (21 Per)	5.03	5.14	A0 p	2.88422	+31	509	2	51	13	+31	31.9
SX Ari (56 Ari)	5.67	5.71	B5-7 IV-V	0.7278925	+26	523	3	6	17	+26	52.8
DO Eri	5.97	6.00	A p	12.448	−12	752	3	50	34	−12	23.5
GS Tau (41 Tau)	5.15	5.22	A0 p	1.227424	+27	633	4	0	28	+27	21.2
IQ Aur	5.35	5.41	B9 Vp	2.4660	+33	1008	5	12	25	+33	38.5
V 592 Mon	6.16	6.31	A2 p	2.976	− 7	1592	6	45	53	− 7	55.4
AX Cam (53 Cam)	5.95	6.08	A2 p	8.015	+60	1105	7	53	10	+60	35.9
BM Cnc (15 Cnc)	5.53	5.65	B9 Vp	4.116	+30	1664	8	6	57	+29	57.4
HV Hya (3 Hya)	5.66	5.76	A2 p	5.57	− 7	2540	8	30	35	− 7	38.3
BI Cnc (49 Cnc)	5.60	5.71	A0 p	5.43	+10	1864	8	39	19	+10	26.6
κ Cnc (76 Cnc)	5.22	5.27	B9 IIIp	5.0035	+11	1984	9	2	20	+11	4.2
CX Leo (45 Leo)	5.97	6.15	B8.5 Vp	—	+10	2152	10	22	22	+10	16.4
GN Com (13 Com)	5.18	5.20	A3 Vp	>1	+26	2344	12	19	17	+26	39.2
AI Com (17 Com)	5.27	5.40	A0 p	5.0808	+26	2354	12	23	55	+26	28.0
UU Com (21 Com)	5.40	5.46	A3 Vp	2.1953	+25	2517	12	26	1	+25	7.2
ε UMa (77 UMa)	1.76	1.79	A0 Vp	5.0887	+56	1627	12	49	38	+56	30.1
α² CVn (12 CVn)	2.78	2.81	A0 IIIp	5.46939	+39	2580	12	51	20	+38	51.5
CW Vir (78 Vir)	4.91	4.99	A2 p	3.7720	+ 4	2764	13	29	4	+ 4	10.2
CR UMa (84 UMa)	5.64	5.67	A2 p	>1	+55	1634	13	42	52	+54	55.9
CU Vir	4.98	5.05	B9 Vp	0.520767	+ 3	2867	14	7	12	+ 2	52.8
CS Vir	5.73	5.93	A3 p	9.2954	−18	3789	14	13	6	−18	15.2
β CrB (3 CrB)	3.65	3.69	A7–F0 IIIp	18.487	+29	2670	15	23	42	+29	27.0
χ Ser (20 Ser)	5.33	5.36	A1 p	1.59584	+13	2982	15	37	5	+13	10.1
V 637 Her (52 Her)	4.78	4.83	A2 p	0.96:	+46	2220	16	46	19	+46	9.4
V 1286 Aql (10 Aql)	5.83	5.93	A4 p	6.05	+13	3838	18	54	11	+13	46.3
V 1288 Aql (21 Aql)	5.06	5.14	B8 II-IIIp	1.7	+ 2	3824	19	8	40	+ 2	7.4
V 1291 Aql	5.61	5.67	A5 p	224.5	− 3	4742	19	48	4	− 3	22.5
AF Dra (73 Dra)	5.16	5.22	A3 IIIp	20.2728	+74	872	20	32	50	+74	36.7
γ Equ (5 Equ)	4.58	4.77	F0 IIIp	314	+ 9	4732	21	5	29	+ 9	43.6
θ¹ Mic	4.77	4.87	A2 p	2.1219	−41	14475	21	14	22	−41	13.9
κ Psc (8 Psc)	4.91	4.95	A2 p	0.5805	+ 0	4998	23	21	49	+ 0	42.4
ι Phe	4.70	4.75	A2 Vp	12.5:	−43	15420	23	29	42	−43	10.1
ET Aqr (108 Aqr)	5.16	5.21	B9 p	3.73	−19	6522	23	46	12	−19	27.9

members are just leaving the main sequence shows that they are very young, and many of them contain fainter stars that are still in the pre-main-sequence stage. Each of them conforms to the main sequence in one range of luminosities.

The range of conditions displayed by these clusters represents an era of relatively slow development. In the next phase the changes proceed more swiftly, and carry the brightest stars into the domain of exceptional luminosity.

The Bubble Reputation—Supergiants

The familiar bright stars of the sky include many hot main-sequence members such as Spica, Regulus, Thuban, Denebola, and Altair. Their masses are moderately high, from two to seven times the sun's. These, as we saw from the Hess diagram (fig. 5.2), are the most numerous stars of their spectral classes because they spend the greater part of their lifetime on or near the main sequence. But many of the most familiar naked-eye stars are very different. They are *supergiants*, distended globes with ten thousand to a hundred thousand times the sun's luminosity and from ten to over a thousand times the sun's diameter. Fine details of their spectra bear the marks of low density and of the corresponding low density and low gravity at their surfaces.

Such supergiants are Alnilam (ϵ Orionis, the "string of pearls"), central star of Orion's belt, Rigel (β Orionis, the "giant's heel"), Canopus (α Carinae), Mirfak (α Persei), Antares (α Scorpii, the "rival of Mars"), and Betelgeuse (α Orionis, the "giant's shoulder"). Three are members of the great Orion complex, a huge association, and Mirfak is a member of an open cluster (table 7.1).

The proportion of double stars is not so great among supergiants as among stars of the main sequence. Both Rigel and Antares have distant companions, hot stars on or near the main sequence. Antares and Betelgeuse are both variable in brightness, with slow and steady changes that show only a suggestion of periodicity of several thousand days. Mirfak seems to be very slightly variable in brightness, a property that is shared by most supergiants. A very few supergiants are even brighter than -7, and are denoted by luminosity class Ia0, but very few stars remain brighter than this for long. Novae and supernovae may do so for a short time during their explosive outbursts.

Table 7.1.   Characteristics of six supergiants.[a]

Formal name	Familiar name	Spectrum	Apparent visual magnitude	Approximate diameter (suns)	Approximate temperature (°K)
ϵ Orionis	Alnilam	B0 Ia	1.70	20	22,000
β Orionis	Rigel	B8 Ia	0.08	40	12,000
α Carinae	Canopus	F0 Ib	−0.73	60	7,500
α Persei	Mirfak	F5 Ib	1.79	80	6,500
α Scorpii	Antares	M1 Ib	1.08 var.	700	2,900
α Orionis	Betelgeuse	M2 Iab	0.80 var.	1,000	2,700

[a] The absolute visual magnitudes of these supergiants are too poorly known to be included in the table; they lie between −4.5 and about −7. Luminosity class I is subdivided into Ia (the brightest) and Ib (the faintest); Iab denotes an intermediate grade.

The stars listed in table 7.1 are among the most luminous known. Alnilam and Rigel must be very massive, from ten to fifty suns or more; Canopus and Mirfak, less luminous, are probably rather less massive. Betelgeuse and Antares, probably with absolute bolometric magnitudes brighter than − 6, could theoretically fall in the range of masses between two and eight times the sun's. On the masses of these two stars we have no direct evidence, but a few comparable red supergiants that are members of binary systems are known to be very massive indeed. The red component of VV Cephei, a supergiant of specturm M2, has a mass at least forty times the sun's, perhaps even more, and the red component of o^2 Cygni (V 1488 Cygni), a K5 supergiant, may have a mass of more than twenty suns. Such stars must be very young. Even a star of five solar masses runs through its distended stage, which marks the end of hydrogen burning in a narrow shell, in less than a million years. The supergiant stage of very massive stars may well last only tens of thousands of years or even less. We do not know of any stars with more than a hundred solar masses. If they exist they must be exceedingly uncommon—and also exceedingly luminous.

Because supergiants must be very young stars, we shall expect that very young clusters will contain them, clusters whose youth is attested by the presence of hot, luminous stars still on the main sequence. In this we are not disappointed, for some of the youngest clusters have supergiant members (table 7.2). Figures 7.1 through 7.11 illustrate such clusters. Each includes at least one supergiant. Several, including the youngest, have two. Ages range from about eight million years for NGC 457 (fig. 7.1) to about twenty million for NGC 6834 (fig. 7.11).

The clusters with the brightest main sequence turnoff have the most

Table 7.2. Positions of thirteen open clusters that contain supergiants. Clusters bright enough to be seen with the naked eye or with binoculars are marked with asterisks.

Name	Position			
	h	m	°	'
*NGC 457	1	12.8	+57	48
NGC 581 (M103)	1	26.6	+60	11
NGC 654	1	37.2	+61	23
*NGC 869 (h Per)	2	12.0	+56	41
*NGC 884 (χ Per)	2	15.4	+56	39
IC 1848	2	43.3	+60	1
*Mel 20 (α Per)	3	15	+48	15
*NGC 2244	6	27.0	+ 4	56
IC 2581	10	23.7	−57	8
NGC 4755 (κ Cru)	12	47.7	−59	48
NGC 6823	19	38.2	+23	4
NGC 6834	19	48.2	+28	9
NGC 7235	22	9.0	+56	47

luminous supergiants. Among those where studies have been carried down to relatively faint magnitudes, several show a scattering of faint stars that have not yet reached the main sequence—IC 2581 (fig. 7.2), NGC 7235 (fig. 7.5), NGC 6823 (fig. 7.6), and IC 1848 (fig. 7.8), for instance. The supergiants, presumably the most massive stars in the respective groups, have moved farthest from the line and are appreciably cooler at the surface than their nearest main-sequence neighbor. They are also cooler than the slightly evolved stars that have begun to swing away. There are no double stars among these supergiants: the B0 star that is marked with a cross in figure 7.2 is the eclipsing star V 348 Carinae, which is not a supergiant.

The supergiant members of the clusters in figures 7.1 through 7.11 are only moderately cool stars, with temperatures upwards of 7000 degrees. But some clusters and associations contain cool supergiants comparable to Betelgeuse and Antares. The very red variable star VY Canis Majoris may be a member of the young cluster NGC 2362 (fig. 6.1). The young Perseus II association, of which the twin clusters h and χ Persei are nuclei, is a veritable nest of such stars. We have already noted that h Persei is the elder brother and χ Persei somewhat younger. Both are rich in blue stars on the main sequence and others that are slightly evolved from it, including the group of β Canis Majoris stars (fig. 7.12). Farther from the clusters but within the association is a group of about twenty red super-

Figure 7.1. NGC 457. The circles denote means for several stars. Isochrones for ten and twenty million years are shown by broken lines. The brightest star is the luminous supergiant φ Cassiopeiae, spectrum F0 Ia; the second brightest is HD 7902, spectrum B6 Ib. The fainter stars that are veering off the main sequence have spectra of luminosity class IV. (Photograph by Harvard Observatory; photometry based on the work of A. F. J. Moffat.)

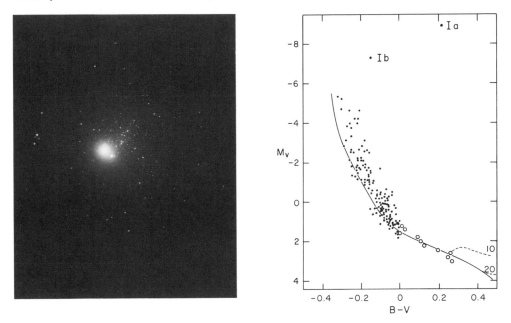

Figure 7.2. IC 2581. This cluster is very like NGC 457. Here the brightest star, HD 90772, is a supergiant of spectrum A7 Ia; the second brightest, HD 90706, has spectrum B2.5 Ib. A cross marks the eclipsing star V 348 Carinae, period 5.5 days. (Photograph by Harvard Observatory.)

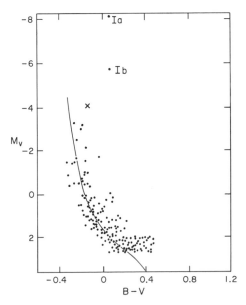

Figure 7.3. NGC 654. The cluster is in the upper part of the photograph. Its brightest star is the F5 Ia supergiant HD 10494; the second brightest has spectrum A0 Ib, so the bright stars are slightly more evolved than in NGC 457. Two other clusters appear in the photograph: NGC 663 to the left below the center, and NGC 659 near the bottom. Both are young clusters. (Photograph by Harvard Observatory.)

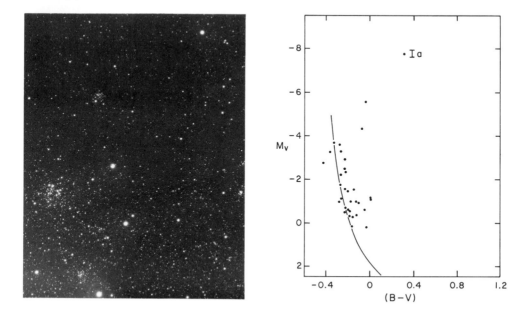

Figure 7.4. NGC 2244. The brightest stars near the main sequence of this very young cluster are blue supergiants; probably the redder star, of luminosity class III, is not a member. The cluster contains a large number of fifteenth to seventeenth magnitude stars that are still approaching the main sequence. (Photograph by Harvard Observatory.)

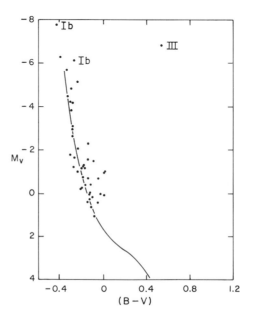

Figure 7.5. NGC 7235. The brightest star has spectrum B9 Iab. Most of the brighter stars to the right of the main sequence have slightly evolved spectra and are in luminosity classes II and III. Circles represent means for fainter stars, many of which have not yet reached the main sequence. Isochrones for one million and three million years are shown and point to an age of about two million years. (Based on the work of A. F. J. Moffat; photograph by the United States Naval Observatory.)

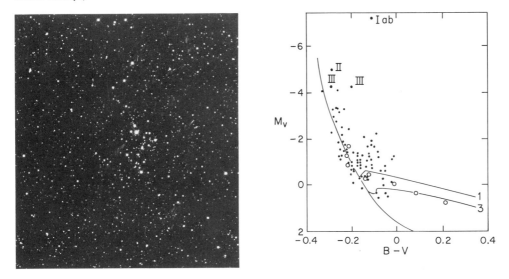

Figure 7.6. NGC 6823. The brightest star has spectrum B0.5 Ib. Circles denote means for fainter stars, most of them approaching the main sequence. Isochrones for one, two, and three million years are shown; they fix the age of the cluster at about two million years. The brightest stars show evidence of dust shells, another index of youth. Two clusters are seen on the photograph, above the center. The one on the right is NGC 6823; the other is NGC 6830. (Based on the work of A. F. J. Moffat.)

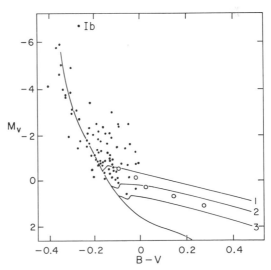

Figure 7.7. NGC 581. The brightest star has spectrum B5 Ib, the second brightest, B2 III. The others with known spectra are main-sequence stars. Broken lines show isochrones for ten million and twenty million years. The means for fainter stars (circles) show that this cluster is older than the preceding ones; nine million years has been suggested. (Diagram based on the work of A. F. J. Moffat; photograph by Harvard Observatory.)

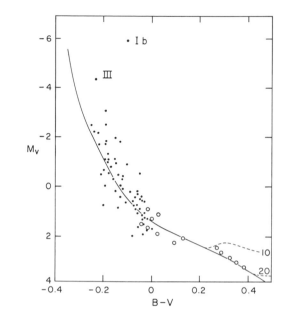

Figure 7.8. IC 1848. The F5 supergiant, marked Ia in the figure, may not be a cluster member. There are a number of pre-main-sequence stars above the lower main sequence, and the brightest main-sequence stars are of classes O and B, so the cluster is very young. (Photograph by the United States Naval Observatory.)

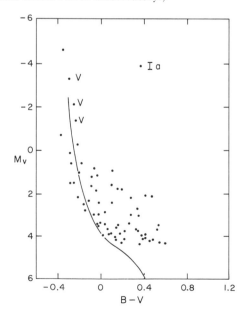

Figure 7.9. NGC 4755 (κ Crucis), at the edge of the heavily obscured region known as the Coal Sack near the Southern Cross. Besides the two blue supergiants, it includes a bright supergiant of class M (outside the limits of the diagram), which has earned the cluster its name of the Jewel Box. Of the two large-scale photographs reproduced, the upper one was made in blue light, the lower in red light. It is easy to pick out the red star, which is relatively much brighter in the lower picture. Below is reproduced a photograph of the Coal Sack on a different scale. The cluster can be seen near the lower edge. Near the middle of the picture is another cluster (NGC 4609), which is obscured by the dark cloud. The very bright stars are α Crucis (to the left) and β Crucis (near the bottom), two of the four members of the Southern Cross. (Photographs by Harvard Observatory.)

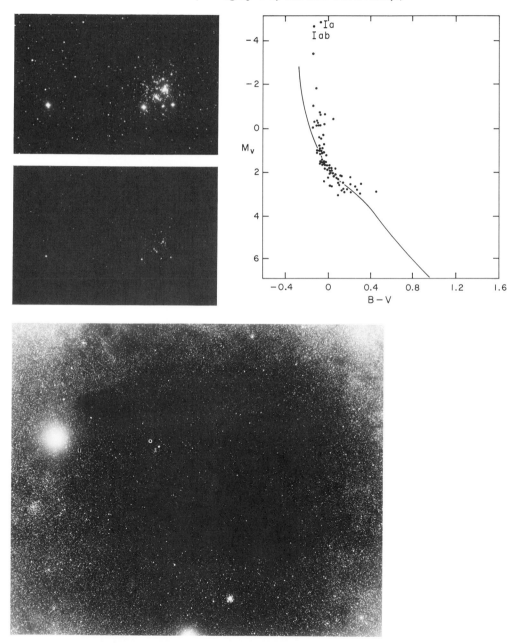

Figure 7.10.   The α Persei cluster. The brightest blue stars are near the main sequence; the brightest star of all, α Persei (Mirfak), is a supergiant of spectral class F5. A circle marks LT Persei, an α Canum Venaticorum star.

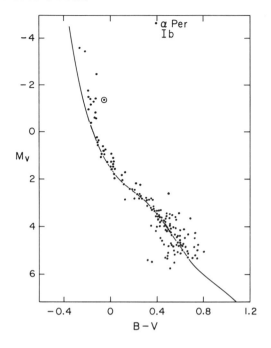

Figure 7.11.   NGC 6834. The brightest star is a supergiant of spectral class F0. The circles represent means of several stars. (Photograph by the United States Naval Observatory.)

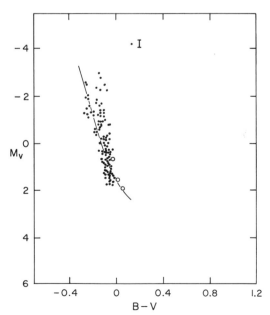

Figure 7.12. Composite color–magnitude array of h and χ Persei (circles, h Persei; dots, χ Persei). Crosses show the red supergiants in the Perseus I association, of which the twin clusters are nuclei. The association also contains at least ten β Canis Majoris variables, slightly evolved stars that occupy a strip just above the brightest two magnitudes of the main sequence. The light lines at the lower end of the main sequence represent the faintest luminosities for stars that have reached the main sequence in χ Persei (above) and in h Persei (below). The broken lines show schematically the isochrones for ten million and twenty million years.

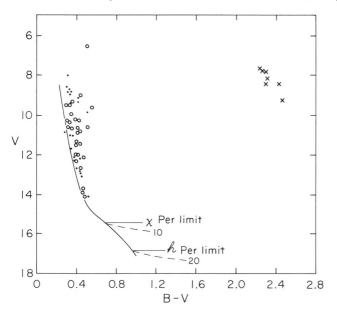

giants. It is the largest such group known and is within one of the most populous stellar communities containing young stars. Many of these red supergiants are known variable stars of small range, and all are probably more or less variable. They must have left the main sequence: the members of the association that are approaching the main sequence are much fainter. Probably they represent the most massive original members of the group, stars that have outrun the bluer supergiants. They are nearing the turning point at which a drastic change of diet will take place and they will begin to consume their internal helium for the first time, burning it into carbon.

Here is a group of stars that seem to be executing some kind of pulsation, with recognizable though ill-defined periodicity. Their periods range from 470 days for SU Persei to 152 days for RS Persei (not counting the even brighter S Persei, where two still longer periods seem to alternate). SU Persei is visually the brighter, RS Persei the faintest. In other words, these stars exhibit roughly a period–luminosity relation, wherein the brightest has the longest period. These are certainly not Mira variables like *o* Ceti, or semiregular variables like Z Aquarii, both of which are far less luminous. Perhaps these variable supergaints are analogues of the red variable stars in the Magellanic Clouds, which are similar in luminosity and display the same sort of relation between luminosity and period. The supergiants in the Perseus association show how a massive star behaves when it has completed its first excursion across the array of stellar properties. As with the Cepheid variables (chapter 8), their periods are evidently governed by their densities, but we do not know in this case what sustains the variations.

The enormous size of these red supergiants, a thousand times that of the sun or even more, is hard to visualize. If the sun were like them, it would engulf the inner planets. The startling dimensions call for independent verification, which is fortunately possible in several ways.

Both Betelgeuse and Antares are large enough and near enough for their angular diameters to be measured with the interferometer. These observations not only verify the enormous sizes but also suggest that their dimensions are variable and that Antares, at least, differs in diameter at different times and in different cross sections. So it may not even be spherical.

Another approach to the sizes of these monster stars in furnished when they are occulted by the moon. The disappearance is not instantaneous, and though the interval must be measured in milliseconds, modern techniques have permitted the measurement of the angular diameters of a number of stars that happen to lie in the path of the moon, including Antares. If the distance is known (inferred, for example, from the luminosity), the linear diameter can be calculated. The enormous sizes of the red supergiants have been amply verified.

For a few stars an even simpler method is available. If the supergiant is a member of a suitably oriented binary system, we can use the eclipses of the two stars by each other to determine their sizes, instead of using eclipses by the moon. Two such eclipsing systems are those known as VV Cephei and BL Telescopii.

A light curve of VV Cephei is shown in figure 7.13. The orbital period is 7430 days, over twenty years. The system consists of a red supergiant, spectrum M2 Ia, and a B star of nearly equal visual brightness. Both spectra are peculiar and show emission lines. From the geometry of the eclipse we find that the diameter of the red star—possibly the largest known star—is 2400 times that of the sun, but the blue star is only 24 times the sun's size. The red star has about forty solar masses, the blue star perhaps fifteen.

Figure 7.13. Variations of light and radial velocity for VV Cephei. (A) Photographic observations. (B) Details of three minima. (C) Visual and photovisual observations. (D) Radial velocity. (From work by Sergei Illarionovich Gaposchkin.)

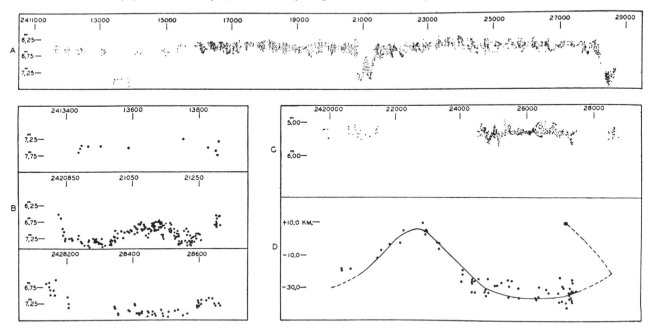

The system of BL Telescopii is much smaller. The orbital period is 778 days, just over two years, and the components are a class F supergiant and a cool star of class M. Here the geometry of the eclipse shows that the components are the same size, about a hundred times that of the sun. The hotter (F8) star is the supergiant, the cooler a low-temperature giant star, neither abnormally bright nor abnormally large. The relative masses are not known, for the motion of the supergiant only has been measured. The M star is too faint for its velocity to have been observed hitherto. We may guess that it is the less massive of the two.

In the system of VV Cephei, then, the more massive red star is the more evolved, but for BL Telescopii the fainter red star seems to have run ahead. Such anomalies are a commonplace among binary systems—the Algol system has already been mentioned—and they will concern us in the final chapter.

After the red supergiant stage we lose sight of the very massive star. What is its later history? If it behaves like less massive stars, and if theory is a valid guide, it will run its later course with ever increasing speed, perhaps toward a grand stellar catastrophe. No star that can be identified as a later stage of a very massive star has been found. The uncommonness of

Figure 7.14.   Distribution of very young open clusters, projected on the plane of the Galaxy. Circles denote associations of pre-main-sequence stars. The arrow points in the direction of the galactic center; the anticenter is in the opposite direction. Short lines are drawn at intervals of one kiloparsec. The galactic center would be about four kiloparsecs below the bottom of the diagram.

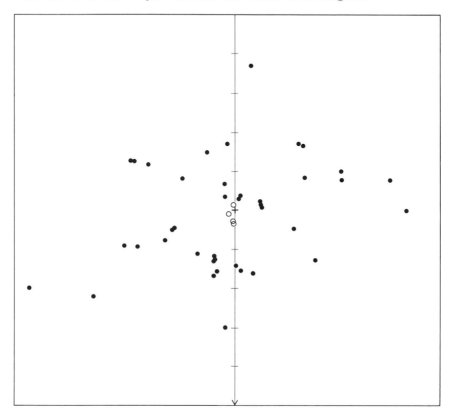

very massive stars, and the probable rapidity of their final stages, would make their sequelae rare and elusive. Later we shall hazard a guess, but for the present we draw a veil.

   Hitherto our examples have been drawn from the youthful clusters, and we may now glance at the part they play in the grand design of the galaxy. Figure 7.14 shows the distribution of the youngest known open clusters, including many that have been used as illustrations of our theme. The diagram covers only part of the Galaxy, for our optical surveys do not extend beyond the galactic center. Evidently these young clusters are not distributed haphazardly, but lie in fairly well-defined strips, which coincide reasonably well (when the uncertainties are considered) with the "spiral" features delineated by the techniques of radio astronomy. The young clusters have been formed in the gaseous substratum, and because of their youth they still lie near their birthplace. On the inner edge of the nearest "spiral arm" lie the Taurus, Chameleon, Ophiuchus, and Corona Austrina dark nebulae, the stellar hatcheries described in chapter 4. Only when they are very near to us are such districts detectable. No doubt similar hotbeds of stellar formation are distributed throughout the variegated spiral structure of our galaxy, along the inner edges of the gaseous clouds. When we look at the distribution of the clusters whose color–magnitude diagrams show them to be older, we shall find that they do not conform to this relatively clean-cut pattern.

Sighing like Furnace—The Cepheid Variable

CHAPTER 8

Returning to the adventures of the star of five solar masses as it pursues its zigzag course across the color–magnitude diagram, we note that the theoretical track makes a turn when helium burning begins in the core (point 7 on fig. 3.3), and yet another when that helium burning reaches major proportions (point 8)—all within a couple of million years. The luminosity of the star increases. It executes a complicated succession of bends and loops across the array of stellar properties, changing all the while in size and surface temperature. When it is moving to the right, its size is increasing and its average density falling; it grows smaller and denser as it moves to the left.

From figure 3.4 it is seen that analogous bends and loops are predicted for stars of about three solar masses and more; the greater the mass, the more pronounced are the loops. For fifteen solar masses the tracks have been followed only as far as the first red giants; subsequent loops, if any, must be traversed very swiftly. The figure contains no information about more massive stars, except to suggest that if they execute similar maneuvers, the time scale must be exceedingly short. This, and the rarity of very massive stars, make it unlikely that the corresponding stages will be readily identifiable, and in fact, they are not.

Because figure 3.4 is the outcome of calculations made with a particular choice of chemical composition and because differences of composition (particularly metal abundance) modify the shape and extent of the bends and loops, a comparison between observation and theory is somewhat tenuous. However, for conditions that correspond to the observed range of composition for open clusters, the predicted paths of stars of over three solar masses will always weave back and forth across the array of stellar properties.

As compared to time spent on the main sequence, the bends and loops are performed very rapidly, so that a given star must pass quickly through the corresponding stages. It is not surprising that few stars are found in these domains of the open clusters. In chapter 7 we met the rare, luminous supergiants, whose temperatures range from that of Alnilam (roughly 22,000 degrees) to that of Mirfak (roughly 6500 degrees) and again from that of Antares to that of Betelgeuse (roughly 2900 and 2700 degrees, respectively). Between these groups, within a range of surface temperature from about 7000 degrees to about 3500 degrees, there are no known supergiants that are not variable in brightness. Indeed, even the stars at the edges of this gap, such as Mirfak, have a slight tendency to variation. Within it, however, all the supergiants show well-marked periodic variation of brightness, surface velocity, and spectrum. These are the *Cepheid variables*, so-called because one of the brightest and first studied is the naked-eye star δ Cephei. The interval of color, spectrum, and surface temperature within which they are found is known as the *Cepheid instability strip*.

Periodic changes in brightness, color, spectrum, and surface velocity furnish evidence that Cepheid variables are pulsating or vibrating bodily. Theoretically they are the best understood of all the intrinsic variable stars. The periods of pulsation are the natural periods (fundamental or overtone) for stars of the corresponding masses and dimensions, and are governed principally by mean density. Stars of the lowest density have the longest periods, and these stars are also the largest and the brightest.

These facts are embodied in the *period–luminosity relation*, one of the most fundamental descriptions of stellar behavior, which is illustrated in figure 8.1. This relation was first detected for the Cepheid variables in the nearby Magellanic Clouds, where the luminosities of a large number of Cepheids, all at nearly the same distance, can readily be seen to be related to their periods. There is every reason to believe that a similar relationship obtains for the Cepheid variables in our own stellar system, and also in the comparatively nearby spiral galaxies in which these variable stars have been studied, such as Messier 31 and Messier 33.

Theory has successfully predicted a period–luminosity relation that conforms to the observed one. The pulsations have been shown to be maintained by a sort of heat engine, which depends for its effectiveness on the ionization and recombination of hydrogen and helium atoms in the outer regions of the star. Although hydrogen has been consumed in

Figure 8.1.   The relation between period and luminosity for the known Cepheid variables in the Large and Small Magellanic Clouds. These stars are chosen because those in each of these systems can be regarded as being at approximately the same distance; a small correction has been made to the magnitudes for the Small Cloud to allow for the fact that it is somewhat more distant than the Large Cloud, and similar stars accordingly appear fainter. There is undoubtedly some real scatter of the points about the mean line; the scatter will have been increased by the thickness and possible tilt of the Clouds, and by observational errors. (Based on the work of Sergei Illarionovich Gaposchkin and Cecilia Payne Gaposchkin, who studied these stars on the large collection of photographs of the Magellanic Clouds that has accumulated since 1890 at the Harvard College Observatory, determined their periods, and obtained their light curves.)

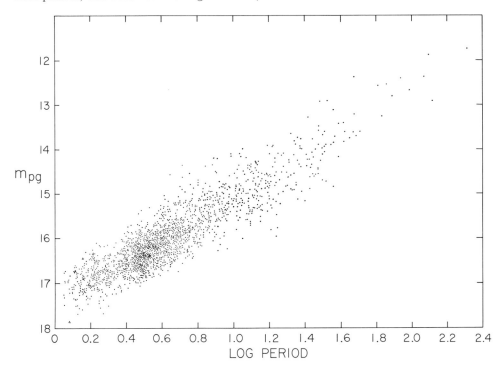

the far interior, and helium is being burned in a thick shell surrounding the stellar core, these atoms still make up the overwhelming majority in the outer envelopes. The pulsating process will operate only within the relatively narrow limits of temperature which will allow the ionization and recombination of these light atoms to act as a pumping mechanism, and these limits span the Cepheid instability strip.

Figures 8.2 through 8.10 show the color–magnitude diagrams of nine open clusters that certainly or probably contain Cepheid variables (see also table 8.1). They are arranged in descending order of the luminosities

Figure 8.2.   NGC 2467 (I Puppis). The Cepheid variable AQ Puppis is indicated by a circle. (Photograph by Harvard College Observatory.)

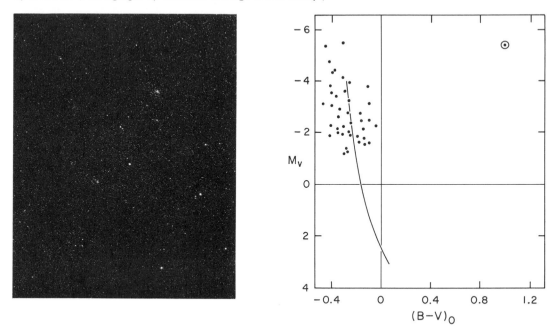

Figure 8.3   Color–magnitude array for the cluster Lynga 6. The Cepheid variable TW Normae is indicated by a circle.

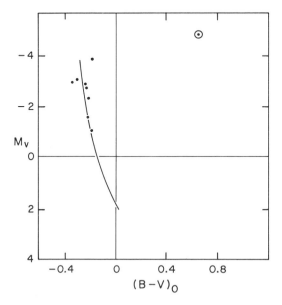

Figure 8.4.   NGC 6087. The Cepheid variable S Normae is indicated by a circle. (Photograph by Harvard Observatory.)

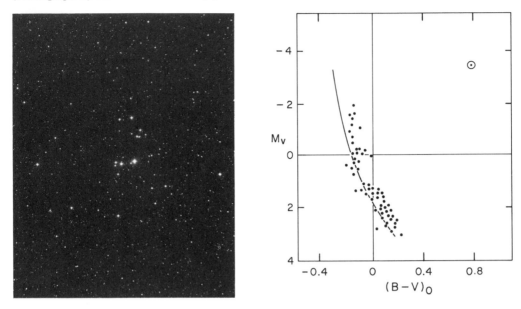

Figure 8.5.   NGC 129. The Cepheid Variable DL Cassiopeiae is indicated by a circle. (Photograph by Harvard Observatory.)

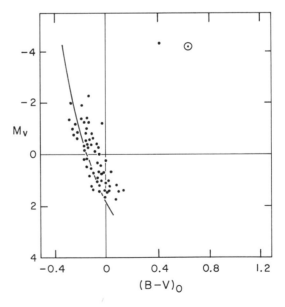

Figure 8.6.   IC 4725 (Messier 25). The Cepheid variable U Sagittarii is indicated by a circle. The cluster is a little to the left of the center of the picture. (Photograph by Harvard Observatory.)

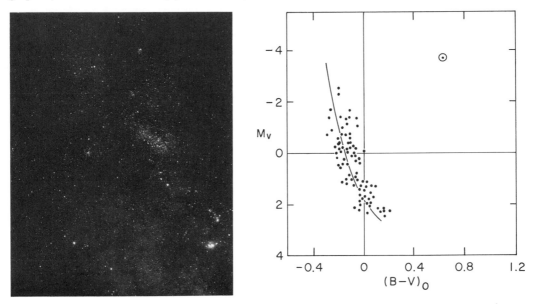

Figure 8.7.   Color–magnitude array for NGC 6649. The Cepheid variable V 367 Scuti is indicated by a circle.

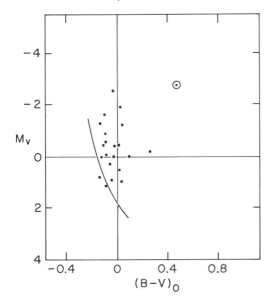

Figure 8.8.   NGC 7790. The Cepheid variables CE Cassiopeiae a, CE Cassiopeiae b, CF Cassiopeiae, and CG Cassiopeiae are indicated by circles. (Photograph by Harvard Observatory.)

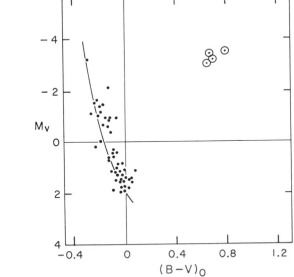

Figure 8.9.   NGC 6067. The Cepheid variable GU Normae is indicated by a circle. (Photograph by Harvard Observatory.)

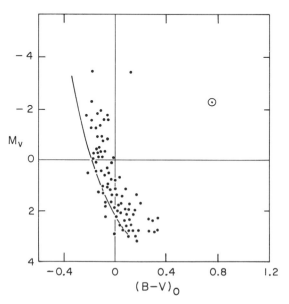

Figure 8.10. Color–magnitude array for NGC 6664. The Cepheid variables Y Scuti and EV Scuti are indicated by circles. Y Scuti (the brighter and redder) is a doubtful member, but EV Scuti definitely belongs to the cluster. A few redder stars are omitted from the diagram.

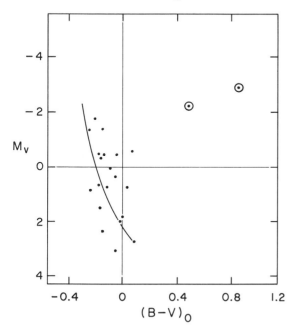

Table 8.1. Positions of nine open clusters that include Cepheid variables.

Name	Position			
	h	m	°	′
NGC 129	0	24.3	+59	40
NGC 2467	7	48.3	−26	8
Lynga 6	16	1.0	−51	47
NGC 6067	16	5.4	−53	57
NGC 6087	16	10.6	−57	39
IC 4725 (M 25)	18	25.8	−19	19
NGC 6649	18	27.9	−10	28
NGC 6664	18	31.3	− 8	18
NGC 7790	23	53.4	+60	39

Figure 8.11. Composite color–magnitude array for the nine open clusters that contain Cepheids. The letters denote respectively AQ Puppis, TW Normae, DL Cassiopeiae, U Sagittarii, S Normae, *a* = CE Cassiopeiae a, *b* = CE Cassiopeiae b, CF Cassiopeiae, CG Cassiopeiae, SZ Scuti, Y Scuti, V 367 Scuti, EV Scuti, GU Normae. The doubtful members are shown by small circles; several δ Scuti stars are indicated by crosses. The main sequence is marked by a line, the Cepheid domain by parallel straight lines. The circled dot is α Persei, a supergiant of class F5 that is slightly variable and may be preparing to become a Cepheid.

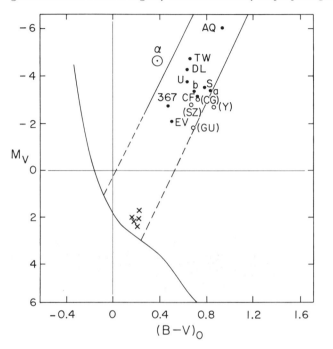

of their Cepheids. Most of them have only one Cepheid, but NGC 7790 contains three, perhaps four. The less certain cluster members are CG Cassiopeiae (NGC 7790, fig. 8.8), GU Normae (NGC 6067, fig. 8.9), and Y Scuti (NGC 6664, fig. 8.10). Figure 8.11 combines the data from all nine clusters and shows how the Cepheids are related to the instability strip and to the main sequence. If they are indeed members of the clusters, these thirteen Cepheids are of determinate luminosity. Figure 8.12 combines them on this basis into a period–luminosity relation, which compares well with figure 8.1.

Intensive studies of Cepheid variables in our own galaxy and in nearby systems—both Magellanic Clouds, Messier 31, Messier 33, and others—show that they have the same gross properties. However, there are important differences from system to system, especially in the way the periods

Figure 8.12. Composite period–luminosity relation for Cepheid variables in open clusters. (1) EV Scuti; (2) GU Normae; (3) CG Cassiopeiae; (4) CE Cassiopeiae b; (5) CF Cassiopeiae; (6) CE Cassiopeiae a; (7) V 367 Scuti; (8) U Sagittarii; (9) DL Cassiopeiae; (10) S Normae; (11) Y Scuti; (12) TW Normae; (13) AQ Puppis.

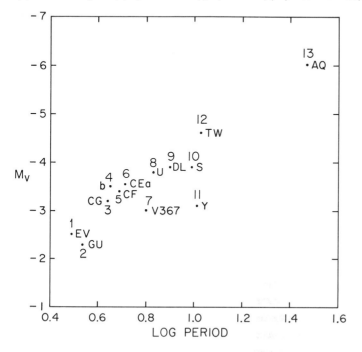

are distributed. Less than five hundred are known in the Galaxy—far fewer than the Mira stars or the RR Lyrae stars that we shall meet later. Probably they play a similar part in the population of Messier 31, but for both Magellanic Clouds they seem to constitute a larger fraction of the stellar community. The possible meaning of these differences will concern us later.

Only about 3 percent of the known Cepheids in our galaxy are members of clusters. This is a far cry from the irregular pre-main-sequence stars, virtually all of which are in clusters or associations. Even for the β Canis Majoris stars, which have not been nearly as completely surveyed, about 40 percent are known to be members of groups. The earlier the stage in the stellar life history, the greater is the observed tendency to be a part of a group.

Many Cepheid variables can be seen with the unaided eye (table 8.2), and a keen observer can perceive changes in brightness from night to night. The type star, δ Cephei, changes visually through nearly a magni-

Table 8.2. Characteristics of nineteen Cepheid variables easily visible to the unaided eye.

Name	Max.	Min.	Spectrum	Period (days)	DM		Position h	m	s	°	'
α UMi (1 UMi)	1.94	2.05	F7 Ib–F8 Ib	3.96978	+88°	8	01	22	33	+88	46.4
β Dor	3.38	4.08	F6 Ia–G2 Iab	9.84200	−62	487	05	32	45	−62	33.3
RT Aur (48 Aur)	5.06	5.83	F5.5–G9 Ib	3.728261	+30	1238	06	22	08	+30	33.3
ζ Gem (43 Gem)	3.68	4.16	F7 Ib–G3 Ib	10.15082	+20	1687	06	58	11	+22	32.1
MY Pup	5.54	5.76	F4 Iab	5.6952	−48	3091	07	35	29	−48	22.4
AH Vel	5.50	5.86	F8	4.22713	−46	3902	08	08	51	−46	20.7
l Car	3.32	4.01	F8–K0 Ib	35.5412	−61	1333	09	42	30	−62	02.8
S Mus	5.90	6.49	F8–G4	9.65885	−69	1646	12	07	24	−69	35.7
R Mus	5.92	6.78	F9–G6	7.50990	−68	1731	12	35	58	−68	51.5
AX Cir	5.64	6.06	F2–G2	5.2734	−63	3436	14	44	28	−63	23.8
X Sgr (3 Sgr)	4.24	4.84	F5–G1	7.01225	−27	11930	17	41	16	−27	47.6
W Sgr	4.30	5.08	F4–G1	7.594710	−29	14447	17	58	38	−29	35.1
Y Sgr	5.40	6.10	F6–G0	5.77335	−18	4926	18	15	30	−18	54.3
FF Aql	5.20	5.55	F5 Ia–F8 Ia	4.470956	+17	3799	18	53	48	+17	13.6
η Aql (55 Aql)	3.50	4.30	F6 Ib–G4 Ib	7.176641	+ 0	4337	19	47	23	+ 0	44.9
S Sge (10 Sge)	5.28	6.04	F6 Ib–G5 Ib	8.382173	+16	4067	19	51	29	+16	22.2
T Vul	5.43	6.09	F5 Ib–G0 Ib	4.435576	+27	3890	20	47	13	+27	52.5
V 1334 Cyg	5.77	5.93	F2 Ib	3.3336	+37	4271	21	15	23	+37	48.9
δ Cep (27 Cep)	3.48	4.34	F5 Ib–G1 Ib	5.366341	+57	2548	22	25	27	+57	54.2

tude and has a period of about five days. The far southern β Doradus has a period of nearly ten days. Both stars are brighter than the fourth visual magnitude at their brightest. Figure 8.13 shows how the brightness, color, and surface velocity of a Cepheid changes during the cycle of variation. The velocities show that the surface of the star rises and falls as the star alternately swells and subsides, and thus provide evidence of bodily pulsation. Because the temperature and the surface brightness are changing at the same time, a Cepheid is not actually brightest when it is biggest. It is smallest during the rise in brightness, largest about halfway down the decline to minimum brightness.

The most familiar and apparently brightest of the Cepheids is the Pole star, α Ursae Minoris. Shakespeare puts into the mouth of the doomed Julius Caesar the proud words, "I am constant as the Northern Star." Ironically, Polaris is one of the least constant of stars. Even its apparent position in the sky is steadily changing as a result of the precessional motion of the earth's axis. It is a visual binary; it is also a spectroscopic binary, moving with a period of nearly thirty years around an unobserved companion; and it is a Cepheid variable of very small range, with a period

Figure 8.13.   Light curves of nineteen typical Cepheid variables, arranged in order of period. Note the gradual change in the form of the curve toward longer periods. The first three are probably pulsating in the first overtone.

No.	Name	Period (days)	No.	Name	Period (days)	No.	Name	Period (days)
1	SU Cassiopeiae	1.95	7	U Sagittarii	6.75	13	Z Lacertae	10.89
2	DT Cygni	2.50	8	η Aquilae	7.18	14	TX Cygni	14.71
3	SZ Tauri	3.15	9	VY Cygni	7.86	15	SZ Cygni	15.11
4	V 386 Cygni	5.24	10	S Sagittae	8.38	16	X Cygni	16.39
5	δ Cephei	5.37	11	BZ Cygni	10.14	17	CD Cygni	17.07
6	MW Cygni	5.96	12	ζ Geminorum	10.15	18	T Monocerotis	27.01
						19	SV Vulpeculae	45.13

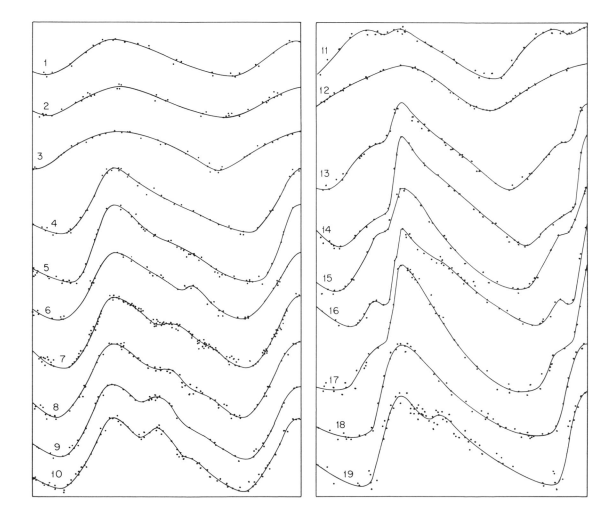

of nearly four days. Those who live in the Northern Hemisphere can look every night at a star that is indeed playing many parts.

Figure 3.4 reminds us that stars with masses greater than about three suns are expected to execute the swift bends and loops that separate the luminous blue stars from the luminous red stars. This region of the color–luminosity array is sparsely populated (a testimony to the speed with which it is traversed). It was Ejnar Hertzsprung who first noted this fact, and the region is named with Hertzsprung gap in his honor. The Cepheid instability strip occupies the middle portion of the gap, which is represented in the Hess diagram by a valley (fig. 5.2). The Hertzsprung gap has a sort of funnel shape and does not extend downward to stars of the sun's luminosity. Its lower end, in fact, coincides with the disappearance of the bends and loops that are predicted for stars with more than three solar masses.

Probably all stars with masses more than three times that of the sun—at least all that are not close binaries—will go through this stage and will therefore become Cepheids. They may in fact pass through the instability strip more than once, growing larger and redder as they move to the right, smaller and bluer as they pass to the left. The consequent changes of density and the associated changes of period are just detectable. One result is that the period–luminosity relation will not be a close correlation, with the stars in the diagram looking like beads on a string. A star of given mass will have different periods at different times but will have almost the same luminosity; therefore the period–luminosity relation will have an appreciable dispersion. The details are complicated by the fact that the star is predicted to traverse different parts of its path with different speeds, and the greater the speed, the shorter is the time taken and the less likely is the corresponding stage to be observed.

It may seem at first that Cepheid variables are not as numerous in open clusters as we might expect. However, the Hertzsprung gap is sparsely populated, and the domain occupied by the instability strip even more so. All the known stars that lie within it in the color–luminosity arrays of open clusters do seem to be Cepheids. Their rarity is a result of the speed with which they move through this phase of their career.

A survey of galactic stars that are not members of clusters leads to similar conclusion. The rarity of Cepheids is matched by the sparse population of the Hertzsprung gap. It seems that the *luminosity function* for Cepheids in our neighborhood (which expresses the relative numbers of stars of different luminosities in a given volume of space) is about what would be expected from the luminosity function of stars with masses greater than three suns, if all such stars become Cepheids at some time in their careers. Both for the stars at large and for the Cepheids, the numbers fall off rapidly toward higher luminosities, as the Hess diagram testifies.

From their situation in the color–magnitude array, all galactic Cepheids thus appear to be less than a hundred million years old. None are found in the very youngest clusters, which suggests that all are at least twenty million years old, counted from the time when they first reached the main sequence. A Cepheid of five solar masses would resemble U Sagittarii, which is in the cluster IC 4725 (fig. 8.6). This star has a period of just under seven days, a typical value for the Cepheids in our immediate vicinity.

According to theory, whether or not a star becomes a Cepheid variable depends on the course of the bends and loops in its tracks of development and the intersection of these bends and loops with the instability strip. Although the details are still obscure, it seems likely that some factor, such as helium content or metal content, can greatly influence the course of the tracks: the loops may be large or small, depending on the composition. If the course of the loops does not differ in different situations, it is hard to understand why the most frequent period for Cepheids is so different in different places.

For Cepheids in our immediate neighborhood, the most frequent period is about seven days. But in the Large Magellanic Cloud (relatively much richer in Cepheids) it is nearer to four days, and in the still richer Small Magellanic Cloud it is less than three days, an almost unknown period for a Cepheid in our vicinity. If galactic and Magellanic Cepheids develop according to the same pattern, the ages of Cepheids in the Large Magellanic Cloud must go back further, and those of the Small Cloud still further, than those of the Cepheids in our own neighborhood. There are many galactic stars older than a hundred million years, but we do not find Cepheids among them, which suggests that the difference is not one of age but of development pattern. A possible explanation is that the Magellanic bends and loops make larger excursions than the galactic ones and can therefore intersect with the instability strip at lower luminosities. It is very tempting to ascribe the difference to composition and its effect on the course of the tracks.

There is, however, strong evidence that the most frequent period for Cepheids is not the same in all parts of the Galaxy. Longer periods predominate nearer to its center, and shorter periods (even less than three days) are almost entirely confined to distant stars in the anticenter region. A similar relation between period and location has been noted in the spiral galaxy Messier 31, in many ways very like our own. Moreover, there is

Table 8.3.   Characteristics of thirty-two δ Scuti stars.

Name	Max.	Min.	Spectrum	Period (days)	DM		Position h	m	s	°	'
β Cas (11 Cas)	2.25	2.31	F2 IV	0.10430	+58°	3	0	3	54	+58	35.7
XX Psc (59 Psc)	5.99	6.03	A5	0.1040	+18	101	0	41	57	+19	1.9
ρ Phe	5.17	5.27	F0n	0.110	−51	209	0	46	8	−51	31.9
AI Scl	5.93	5.98	A7 III	0.05	−38	420	1	8	9	−38	23.2
VX Psc (97 Psc)	5.90	5.92	A4 III	0.131	+17	210	1	24	29	+17	50.3
UV Ari (38 Ari)	5.18	5.22	A7 IV	0.06:	+11	377	2	39	31	+12	1.4
V 376 Per	5.8	5.91	A9 IV	0.097, 0.067	+43	818	3	42	14	+43	39.3
IM Tau (44 Tau)	5.37	5.44	F2 IV	0.144923	+26	686	4	4	44	+26	13.2
o¹ Eri (38 Eri)	4.00	4.05	F2 II–III	0.0815	− 7	764	4	6	59	− 7	5.8
V 483 Tau (57 Tau)	5.56	5.59	F0 V	0.054	+13	663	4	14	21	+13	47.6
V 696 Tau (58 Tau)	5.20	5.26	A9 V	0.036	+14	682	4	14	56	+14	51.3
V 480 Tau (97 Tau)	5.09	5.11	A7 IV–V	0.042	+18	743	4	45	32	+18	40.2
KW Aur (14 Aur)	4.94	5.10	A9 IV	0.08748	+32	922	5	8	53	+32	34.3
V 1004 Ori (59 Ori)	5.88	5.89	A5m	0.054:	+ 1	1171	5	53	13	+ 1	49.6
V 474 Mon (1 Mon)	5.93	6.36	F2 IV	0.13494	− 9	1284	5	54	16	− 9	23.3
ρ Pup (15 Pup)	2.68	2	F6 IIp	0.14088	−23	6828	8	3	17	−24	0.9
FZ Vel	5.15	5.17	F0 III	0.065	−46	4810	8	55	28	−46	50.8
υ UMa (29 UMa)	3.77	3.86	F2 IV	0.133	+59	1268	9	43	51	+59	30.4
AI CVn (4 CVn)	5.97	6.15	F3 IV	0.139	+43	2218	12	18	52	+43	5.8
FM Vir (32 Vir)	5.20	5.23	A5m	0.07	+ 8	2639	12	40	34	+ 8	13.2
AO CVn (20 CVn)	4.70	4.75	F3 IIIp	0.135	+41	2380	13	13	4	+41	5.9
κ² Boo (17 Boo)	4.50	4.54	A8 IV	0.06682	+52	1782	14	9	54	+52	15.4
γ Boo (27 Boo)	3.23	3.28	A7 III	0.290:	+38	2565	14	28	3	+38	44.9
δ Ser (13 Ser)	4.20	4.25	F0 IV	0.134	+11	2821	15	30	1	+10	52.3
γ CrB (8 CrB)	3.80	3.86	A0 III–IV	0.030	+26	2722	15	38	33	+26	36.7
CL Dra	4.95	5.00	F0 IV	0.063	+55	1793	15	55	24	+55	2.0
o Ser (56 Ser)	4.20	4.26	A2 IV–V	0.053	−12	4808	17	35	47	−12	49.4
δ Sct	4.98	5.16	F3 III–IV	0.193770	− 9	4796	18	36	48	− 9	8.9
ρ¹ Sgr (44 Sgr)	3.90	3.93	F0 IV–V	0.050	−18	5322	19	15	52	−18	2.1
δ Del (11 Del)	4.39	4.49	F0 IVp	0.134, 0.158	+14	4403	20	38	47	+14	42.9
τ Cyg (65 Cyg)	3.65	3.84	F2 III–IV	0.143:	+37	4240	21	10	48	+37	37.1
τ Peg (62 Peg)	4.60	4.62	A5 IV–V	0.05433	+22	4810	23	15	41	+23	11.6

growing evidence of a systematic change in chemical composition from the center to the edge of such a galaxy. This may well be related to the differences of period frequency; if so, it fortifies the suggestion that composition lies at the bottom of the observed contrast between galactic and Magellanic Cepheids.

Another glance at figure 3.4 raises a further question. If we pursue the course of the instability strip to lower luminosities, we find that (for the adopted composition) it no longer intersects loops in the tracks for stars with less than three solar masses. It reaches the predicted main sequence for masses about one and one-half times the sun's, and jsut above this it intersects the small crotchet that we have associated at higher luminosities and surface temperatures with the β Canis Majoris and perhaps with the α Canum Venaticorum stars. At this juncture we find another group of variable stars of very short period, the so-called δ Scuti stars.

There are reasons for surmising that these stars may bear some relationship to the Cepheid variables. They display a similar correlation between period and spectrum and even conform roughly to the period–luminosity curve when extrapolated to very short periods (table 8.3). Two such stars have been found in Praesepe, one in the Coma Berenices cluster, one in the Hyades, possibly two in the Pleiades. Figure 8.11 shows their relationship to the main sequence and to the instability strip. These stars are intrinsically much fainter than Cepheids, and their variations are so small and rapid that they are hard to study. Ten percent of the known δ Scuti stars are members of clusters, a far higher proportion than for Cepheids–an indication of slower development.

We saw in the last chapter that the clusters containing supergiants follow the spiral structure of the central plane of the Galaxy. The clusters that contain Cepheids are not so young, and the pattern defined by their distribution is less clean-cut. In any case, they are not numerous. Many Cepheids are not members of clusters, but we may suppose that their ages and histories are otherwise similar. It is interesting to see whether their distribution conforms to that of the clusters. Those of longest period, which must be the most luminous and the youngest, do indeed show indications of spiral structure, but the less luminous ones cannot qualify as spiral indicators.

Many studies of the galactic distribution of Cepheid variables have been made, and though they differ in detail (largely because of the difficulty of allowing for the effects of obscuration), they all paint the same general picture. The Cepheids of long period show definite structure; those with periods less than about ten days do not. Figure 8.14, the outcome of my own census of Cepheids, shows many Cepheids of long period, and few of short period, toward the galactic center, and many short-period Cepheids, but few of long period, toward the anticenter. The former effect could result from incompleteness of the survey in highly obscured regions, but the latter is undoubtedly real. The extraordinary richness in Cepheid variables of the Carina region (galactic longitude about 290°), with a great variety of periods and ages, is very striking; so is its richness in very young clusters and its comparative poverty in really old inhabitants.

Figure 8.14. Distribution of Cepheid variables in our part of the Galaxy. The cross indicates our position; galactic longitudes are marked at the edges, 0 indicating the *direction* of the galactic center (the diagram extends only about halfway to the center itself). Dots indicate Cepheid variables, projected on the galactic plane. Few are actually far from it. Large dots indicate those Cepheids with periods over ten days. For comparison, circles indicate the positions of emission nebulae, which lie in or near the galactic plane and define its "spiral structure" more clearly than the Cepheids do.

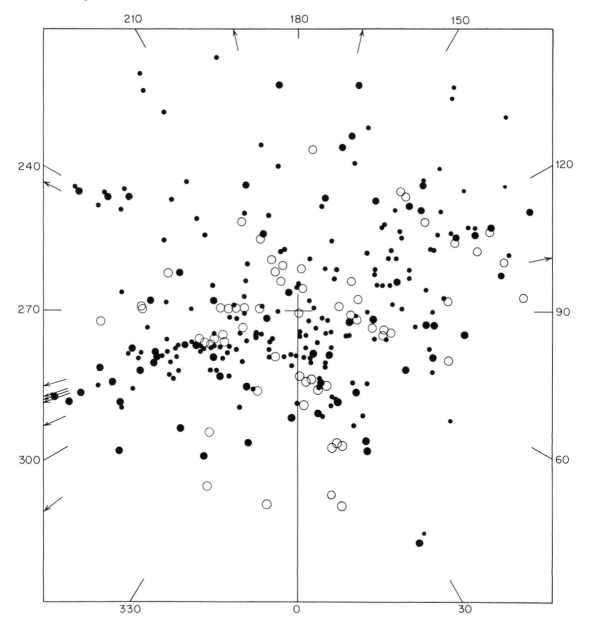

What happens to a star after it has played its part as a Cepheid? In its repeated excursions across the color–magnitude domain it has more than once been a red star and has accordingly been much distended. The low surface gravity that goes with large size will have encouraged loss of material from its surface. After the final loop the temperature falls even further than at earlier redward excursions and mass loss becomes even more probable. An appreciable loss of mass must shift the star to a lower track, with consequences still incompletely explored. In any case, after executing its loops and bends it becomes a distended, cool star—a *red giant* or supergiant.

The Red Giant

CHAPTER 9

Few if any galactic Cepheids are more than a hundred million years old. As the stellar community ages and more stars move away from the main sequence, older and older stars reach giant dimensions beyond the instability strip. At least in the open clusters of our own neighborhood, stars with one to three solar masses appear to pass uneventfully, without executing bends and loops, from the post-main-sequence crotchet to the first distended stage where helium burning begins in earnest. The theoretical diagram does not show a marked crotchet for stars of one solar mass and less. Moreover, for these masses the tracks tend to converge. Whereas stars of high mass and luminosity move nearly horizontally across the color–luminosity diagram, for masses less than two suns the tracks rise more and more steeply. Thus, while supergiants are clearly separated from bright giants, red giants crowd together more and more in luminosity, producing the minor peak evident in the Hess diagram.

We have already seen the group of red supergiants in the Perseus II association, all more or less variable. Among the apparently brightest of such stars is the erratically variable μ Cephei, one of the most luminous stars known, which fluctuates slowly between the third and fifth magnitudes. Fainter than these supergiants, but still brighter than the red giants, are the so-called "bright giants," distinguished from the former and the latter by spectroscopic detail and described by luminosity class II. Three well-known ones are listed in table 9.1. All are double or multiple stars.

Dubhe is a visual double star with a companion rather like the sun; the orbital period is about 44 years. Albireo has already appeared in our list of main-sequence stars as the companion of a B8 V star. It is itself a closer double, as is shown by its composite spectrum. Ras Algethi is a member

Table 9.1. Characteristics of three red bright giants.

Formal name	Familiar name	Spectrum	Apparent visual magnitude	Absolute visual magnitude[a]	Approximate radius (suns)
α Ursae Majoris	Dubhe	K0 II–III	1.79	−1.0	90
β Cygni	Albireo	K5 II	3.42	−2.3	110
α Herculis	Ras Algethi	M5 II	3.1	−2.4	300

[a] Average for the spectral class.

of a triple system; the other component, which shares its motion, is a spectroscopic binary—a giant G star and a giant F star with an orbital period of 51.59 days. Ras Algethi itself is a semiregular variable, as are many other stars of similar spectrum. The whole system is in fact surrounded by an expanding envelope of gas. This red star is observed to be spilling off material into space at a rate of about 3×10^{-8} solar masses (about 6×10^{19} tons) a year. If, as we have every reason to believe, this is typical behavior for a luminous red star, the effect on its subsequent career must be great.

About two magnitudes fainter than these bright giants are the red giants proper, stars of spectra G, K, and M of luminosity class III. Such stars are quite common in the galactic field, and many are well known naked-eye stars. Table 9.2 lists ten of them, of which four are known to be

Table 9.2. Characteristics of ten red giants.

Formal name	Familiar name	Spectrum	Apparent visual magnitude	Absolute visual magnitude[a]
α Aurigae	Capella	G4 III, G0 III?	0.09	+0.6, +0.3
τ Persei	—	G5 III	3.09	+0.3
α Phoenicis	—	K0 III	2.39	+0.2
β Geminorum	Pollux	K0 III	1.15	+0.2
α Arietis	—	K2 III	2.00	−0.1
α Boötis	Arcturus	K2 IIIp	0.06	−0.1?
α Hydrae	Alphard	K4 III	1.99	−0.2
α Tauri	Aldebaran	K5 III	0.86	−0.3
β Andromedae	Mirach	M0 III	2.03	−0.4
α Ceti	Menkar	M2 III	2.52	−0.5

[a] Average for the spectral class.

double or multiple. Sizes range from about ten times to about a hundred times that of the sun.

Seven of the ten stars, we notice, are the brightest in their respective constellations, a testimony to their relative commonness among naked-eye stars. Their affiliations are various. Capella consists of two red giants with an orbital period of 105 days; there are also two much fainter dwarf companions. Another spectroscopic binary is τ Persei, but here the companion is a main-sequence A star. The system of α Phoenicis, another spectroscopic binary, has a period of 3849 days, but the second star has not been observed; it must be much bluer than the visible one. Aldebaran has a companion, a main-sequence M star. Though it seems to lie within the Hyades, this bright star is not known to be a member either of the cluster or the associated group.

Arcturus is an especially interesting star. The peculiarity of its spectrum consists in the exceptional weakness of the metallic lines. The star must actually be poor in metals—the first such star that we have encountered. Besides this, the star has an unusual motion that does not conform to the rotation of the Galaxy. As we proceed to our study of the most evolved clusters, we shall find many stars that are poor in metals, some very much poorer than Arcturus. The difference must point to a different history, and should warn us not to draw conclusions hastily about the age or mass of this star and others that show similar peculiarities. Arcturus is the (apparently) brightest star with these properties, and as such has been intensively studied.

Having made the acquaintance of a number of well-known red giants, we turn to a group of familiar clusters that illustrate this stage of stellar development (figs. 9.1 through 9.5; table 9.3). They range from NGC 6475 (Messier 7), aged about one hundred and ten million years, to Coma Berenices, aged nearly a thousand million years and very similar to the Hyades. All of them contain at least one red giant; the most populous—NGC 6705 (Messier 11) and NGC 2099 (Messier 37)—have many. The youngest, Messier 7, has only one, but this star is itself a spectroscopic binary, one of the unusually large number of binaries in the cluster.

Even the youngest of this group of clusters has turned off its main sequence at too low a luminosity to be involved in the instability strip: they contain no Cepheids, and their brightest stars do not exceed two solar masses. At most these clusters display the δ Scuti stars that we have suggested to be associated with the crotchet just above the main sequence; such stars are indeed known in Praesepe and Coma Berenices. At these masses the tracks for red-giant stars have begun to converge, as shown in figure 3.4. Instead of the excitement of pulsation, the stars make their

Figure 9.1.   NGC 6475 (Messier 7). The cluster is rich in binary stars, of which four are marked with circles. (Photograph by Harvard Observatory.)

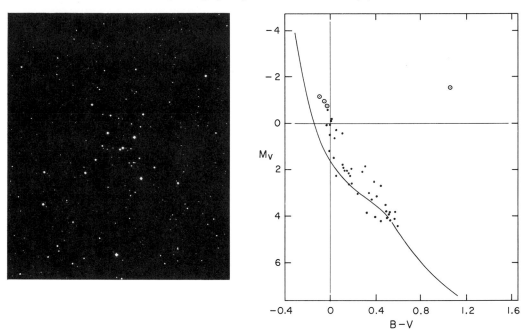

Figure 9.2.   NGC 6705 (Messier 11). There is a well-developed giant branch. One of the brighter blue members is the eclipsing binary BS Scuti, with a spectrum of class A7. (Photograph by Harvard Observatory.)

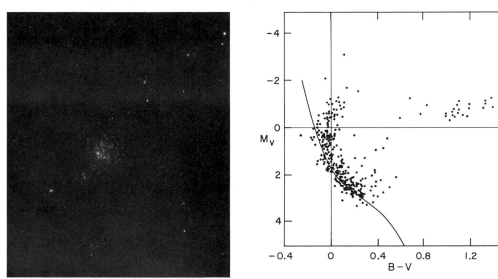

Figure 9.3.   NGC 2099 (Messier 37). The cluster is at the left edge of the photograph; near the center is the older cluster NGC 1960 (Messier 36), farther to the right is the still older NGC 1912 (Messier 37). (Photograph by Harvard Observatory.)

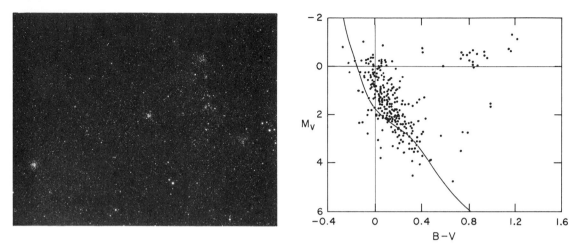

Figure 9.4.   NGC 2632 (Praesepe), with a few red giants and at least one anomalous blue star. The cluster contains several interesting stars, including the eclipsing binary TX Cancri and the "metallic-line star" ϵ Cancri. There are also a number of seventh and eighth magnitude δ Scuti stars in the cluster, and at the sixteenth to eighteenth magnitudes, a number of pre-main-sequence stars. (Photograph by Harvard Observatory.)

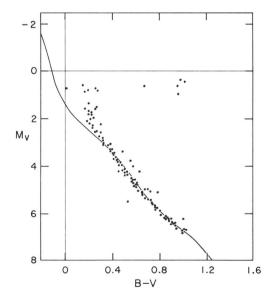

Figure 9.5. The cluster Coma Berenices (Melotte 111). Circled dots mark two α Canum Venaticorum stars, AI Comae and UU Comae. AI Comae is the brighter. A cross marks the δ Scuti star FM Comae; another, somewhat fainter, is GM Comae. Several much fainter variables, below the limit of the diagram, are pre-main-sequence stars such as CV and FO Comae. (Photography by Harvard Observatory.)

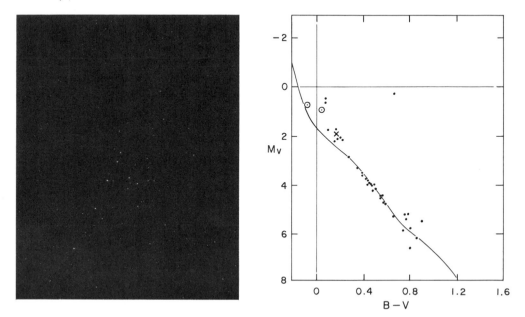

Table 9.3. Positions of six open clusters that contain red giants. All are bright enough to be seen with the naked eye or with binoculars.

Name	Position			
	h	m	°	'
Hyades	4	21	+15	37
NGC 2099 (M37)	5	45.8	+32	31
Praesepe (M44)	8	34.3	+20	20
Coma Berenices	12	20.0	+26	40
NGC 6475 (M7)	17	47.5	−34	47
NGC 6705 (M11)	18	45.7	− 6	23

transition to a brief phase of respectable tranquillity as red giants. They extend their excursions to the red, until at a certain point (near where the spectrum passes from K to M and begins to display molecular absorptions) they develop a tendency to vary. These variations are most marked when the star is reddest and also when it is most luminous. Ras Algethi has already reached this stage and has become a semiregular variable.

Table 9.4. Characteristics of forty-five irregular red variables.

Name	Max.	Min.	Spectrum	DM		Position h	m	s	°	'
GIANTS										
η Scl	4.7	4.90	M4	−33°	152	0	22	58	−33	33.6
BQ Tuc	5.60	5.80	M5	−63	83	0	49	29	−63	24.9
WW Psc	5.95	6.12	M2	+ 5	131	0	54	39	+ 5	56.6
ψ Phe	4.3	4.5	M4	−46	552	1	49	38	−46	47.6
τ⁴ Eri (16 Eri)	3.59	3.72	M3	−22	584	3	15	4	−22	7.3
π Eri (26 Eri)	4.38	4.44	M2	−12	707	3	41	25	−12	24.9
γ Ret	4.42	4.64	M5	−62	312	3	59	27	−62	26.3
α Tau (87 Tau)	0.75	0.95	K5	+16	629	4	30	11	+16	18.3
RX Lep	5.68	7.0	M6	−12	1092	5	6	43	−11	58.4
SW Col	5.71	6.05	M1	−39	1940	5	20	6	−39	46.3
UW Lyn (1 Lyn)	4.93	5.04	M3	+61	869	6	8	42	+61	32.9
μ Gem (13 Gem)	2.76	3.02	M3	+22	1304	6	16	55	+22	33.8
NP Gem	5.80	6.02	M1	+17	1479	6	56	37	+17	53.9
CG UMa	5.87	5.95	M4	+57	1214	9	14	23	+57	7.4
VY UMa	5.89	6.5	C6	+68	617	10	38	7	+67	56.1
VY Leo (56 Leo)	5.69	6.03	M5	+ 6	2369	10	50	50	+ 6	43.1
CO UMa	5.79	5.95	M5	+37	2162	11	3	49	+36	51.1
ω Vir (1 Vir)	5.23	5.37	M4.5	+ 8	2532	11	33	18	+ 8	41.3
μ Mus	4.6	4.8	K3	−66	1649	11	43	26	−66	15.5
CQ Dra (4 Dra)	4.95	5.04	M3	+70	700	12	25	44	+69	45.3
ψ Vir (40 Vir)	4.7	4.8	M3	− 8	3449	12	49	9	− 8	59.8
BY Boo	5.1	5.28	M4	+44	2325	14	3	56	+44	19.8
FL Ser	5.79	6.03	M4	+19	2935	15	7	31	+19	21.2
δ¹ Aps	4.66	4.87	M4	−78	1092	16	5	24	−78	26.8
V 636 Her	5.83	6.03	M4	+42	2749	16	44	8	+42	25.0
η Sgr	3.08	3.12	M3	−36	12423	18	10	51	−36	47.6
V 3872 Sgr (62 Sgr)	4.45	4.76	M4	−28	16355	19	56	31	−27	59.3
EN Aqr (3 Aqr)	4.41	4.45	M3	− 5	5378	20	42	28	− 5	23.6
T Cyg	5.0	5.5	K3	+33	4028	20	43	11	+34	0.4
MO Cep (18 Cep)	5.28	5.37	M5	+62	2028	22	0	53	+62	38.0
λ Aqr (73 Aqr)	3.70	3.80	M2.5	− 8	5968	22	47	24	− 8	6.7
β Peg (53 Peg)	2.31	2.74	M2	+27	4480	22	58	56	+27	32.5
GZ Peg (57 Peg)	5.0	5.16	M4	+ 7	4981	23	4	29	+ 8	8.1
χ Aqr (92 Aqr)	4.9	5.25	M3	− 8	6076	23	11	40	− 8	16.3
HH Peg (80 Peg)	5.74	6.0	M3	+ 8	5127	23	46	15	+ 8	45.5
XZ Psc	5.55	5.57	M5	− 0	4585	23	49	39	− 0	26.8
YY Psc (30 Psc)	4.35	4.41	M3	− 6	6345	23	56	50	− 6	34.2
SUPERGIANTS										
π Aur (35 Aur)	4.24	4.34	M3.5	+45	1217	5	52	31	+45	55.7
o¹ CMa (16 CMa)	3.78	3.99	K3	−24	4567	6	49	59	−24	3.5
σ CMa (22 CMa)	3.43	3.49	M0	−27	3544	6	57	44	−27	47.5
MZ Pup	5.2	5.44	M1	−32	4796	8	0	22	−32	23.5
NS Pup	4.4	4.5	K3	−39	4084	8	7	47	−39	19.2
λ Vel	2.14	2.22	K4	−42	4990	9	4	19	−43	1.7
V 337 Car	3.36	3.42	K5	−60	1817	10	13	44	−60	50.0
β Gru	2.0	2.3	M3–4	−47	14308	22	36	41	−47	24.5

Table 9.5. Characteristics of thirty-one semiregular red variables.

Name	Max.	Min.	Period	Spectrum	DM		Position h	m	s	°	′
GIANTS											
TV Psc (47 Psc)	4.65	5.42	49±	M3	+17°	55	0	22	50	+17	20.4
ρ Per (25 Per)	3.30	4.0	50±	M4	+38	630	2	58	46	+38	27.1
TW Hor	5.25	5.95	158±	C7	−57	513	3	10	2	−57	41.9
R Dor	4.8	6.6	338±	M7	−62	372	4	35	35	−62	16.5
DM Eri (54 Eri)	4.28	4.36	30	M4	−19	988	4	36	4	−19	51.9
o¹ Ori (4 Ori)	4.65	4.88	30±	M2(S)	+14	777	4	46	52	+14	5.0
WZ Dor	5.1	5.28	40	M4	−63	420	5	6	47	−63	31.6
L₂ Pup	2.6	6.2	140.4	M5–M6	−44	3227	7	10	30	−44	28.3
BP Cnc (27 Cnc)	5.41	5.75	40±	M3	+13	1912	8	21	12	+12	59.0
RX LMi	5.98	6.16	150±	M4	+32	2066	10	36	35	+32	13.2
V 763 Cen	5.55	5.80	60±	M3	−46	7199	11	30	23	−46	49.2
II Hya	4.85	5.12	61	M4	−26	8789	11	43	42	−26	11.6
ε Mus	4.95	5.36	55±	M6	−67	1931	12	12	10	−67	24.3
Y CVn	8.2[a]	10.0[a]	158.0	C5	+46	1817	12	40	26	+45	59.2
FS Com (40 Com)	5.5	6.1	58:	M5	+23	2538	13	1	31	+23	9.1
V 744 Cen	5.78	6.55	90±	M8	−49	8095	13	33	47	−49	26.5
ET Vir	4.80	5.00	80	M3	−15	3817	14	5	23	−15	49.8
W Boo (34 Boo)	4.7	5.4	30±	M3	+27	2413	14	39	2	+26	57.2
V 768 Cen	5.93	6.15	60 to 80	M3	−36	9645	14	42	26	−36	13.0
σ Lib (20 Lib)	3.20	3.36	20	M3.5	−24	11834	14	58	13	−24	53.4
g Her (30 Her)	5.7	7.2	70±	M6	+42	2714	16	25	22	+42	6.1
R Lyr (13 Lyr)	3.88	5.00	46.0	M5	+43	3117	18	52	18	+43	48.9
NU Pav	4.91	5.26	80±	M6	−59	7564	19	53	19	−59	38.9
EU Del	5.84	6.9	59.5	M6	+17	4370	20	33	21	+17	55.1
AG Cap (47 Cap)	5.90	6.14	25±	M3	− 9	5833	21	40	56	− 9	44.2
ε Oct	4.96	5.36	55±	M6	−81	995	22	8	51	−80	56.3
π¹ Gru	5.41	6.70	150±	S4	−46	14292	22	16	37	−46	27.1
SUPERGIANTS											
α Ori (58 Ori)	0.42	1.3	2070	M2	+ 7	1055	5	49	45	+ 7	23.3
α Sco (21 Sco)	0.88	1.80	1733	M1–2	−26	11359	16	23	16	−26	12.6
α Her (64 Her)	3.0	4.0	6 yrs?	M5	+14	3207	17	10	5	+14	30.3
μ Cep	3.6	5.1	700–900	M2	+58	2316	21	40	27	+58	19.3

[a] Photographic magnitudes.

Red variable stars are of many kinds (tables 9.4 and 9.5). We have already met the variable supergiants in the Perseus association, which are barely periodic. The same is true of Ras Algethi and a host of similar stars, all of which undergo but small changes of brightness. If the red giants in figures 9.1 through 9.5 are variable, their changes of brightness are very small.

There are many red variables with more conspicuous behavior. Such is CO Cygni, a giant K star that varies slowly and irregularly through about a magnitude. More striking is a star like Z Aquarii, a giant M star that changes visually through about 2.5 magnitudes with a period of 136 days. But this is not an ordinary M star, for its spectrum shows bright hydrogen lines. It points the way to the more spectacular, better-known Mira stars, which may vary visually through as much as ten magnitudes.

But if we wish to place Mira stars in the general pattern of stellar development by finding them in open clusters, we look in vain. Indeed, open clusters display very few red variables with pronounced changes of brightness. The percentage of Cepheids that are members of open clusters was small, but for Mira stars it is zero. In line with our findings about other types of variable stars, this argues a greater age for Mira stars than for any known open cluster, or a different pattern of development. The blame might be laid on a very short lifetime, but Mira stars are so numerous as to make this implausible. If we are to place them in the pattern of stellar development, we must look elsewhere than in the range of conditions presented by the open clusters.

The Oldest Open Clusters

The stellar community ages. Stars move away from the main sequence, stripping it down more and more, and the form of the color–magnitude diagram changes. The red giants of the Hyades, Coma Berenices, and Praesepe have about the same visual magnitude as they had when they left the main sequence. Less massive stars grow more luminous as they leave the main sequence and become so-called subgiants before they arrive at the giant tip. The oldest known open clusters all have shorn-off main sequences; the color–magnitude diagram twists off through the subgiants, then rises toward the giant tip.

Seven old open clusters are shown in figures 10.1 through 10.7 (see also table 10.1). They are arranged in order of age, from about five hundred million years for NGC 2281 to about eleven thousand million for NGC 188, and illustrate the slow changes that the pattern undergoes over this long interval.

The Hertzsprung gap is evident in the five youngest, and becomes progressively narrower in the older ones. It is absent from NGC 2682 (Messier 67) and NGC 188 (figs. 10.6 and 10.7). The crowding of the red giants (the little "hill" in the Hess diagram) is noticeable for NGC 6940, NGC 2477, and NGC 7789 (figs. 10.2, 10.3, and 10.4). The ascent to the red-giant tip grows progressively steeper with advancing age (fig. 10.8).

Most of these diagrams show a scattering of stars that lie outside the main pattern. Partly this blurred pattern is a result of observational uncertainty. It is least conspicuous for NGC 752 (fig. 10.5) and NGC 2281 (fig. 10.1), the two nearest to us, and is especially noticeable for NGC 7789, the one most distant from us (fig. 10.4). Again we must recall that cluster membership cannot always be assigned with certainty, even for

Figure 10.1.   NGC 2281, which includes three red giants. (Photograph by Harvard Observatory.)

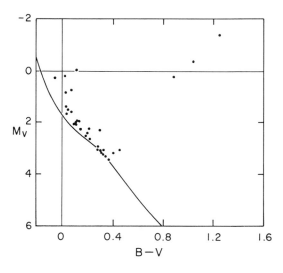

Figure 10.2.   NGC 6940. There are a number of red giants, and the cross marks the semiregular red variable FG Vulpeculae, a probable member. The brighter stars near the top of the main sequence are so-called blue stragglers, which pose an interesting problem. (Photograph by the United States Naval Observatory.)

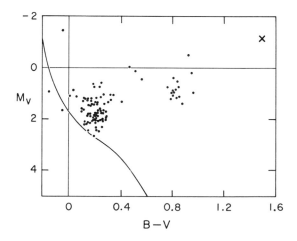

Figure 10.3. NGC 2477. There are a large number of red-giant stars, possibly also a few blue stragglers lingering near the main sequence. (Photograph by Harvard Observatory.)

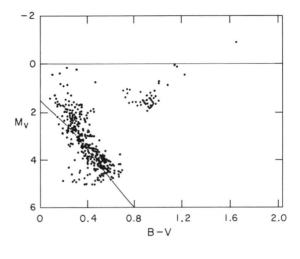

Figure 10.4. NGC 7789. (Photograph by Harvard Observatory.)

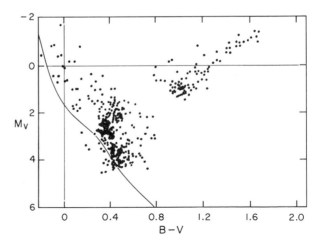

Figure 10.5. NGC 752. This very sparsely populated cluster has a large representation of red giants. There is also one relatively bright star near the main sequence. The circle denotes the eclipsing star DS Andromedac. (Photograph by Harvard Observatory.)

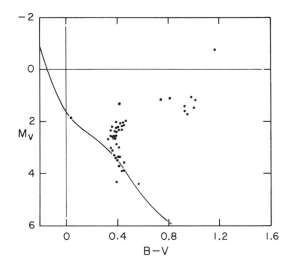

Figure 10.6. NGC 2682 (Messier 67). There is a well-developed red-giant branch, but at least some of the blue stragglers are really members of the cluster. (Photograph by Harvard Observatory.)

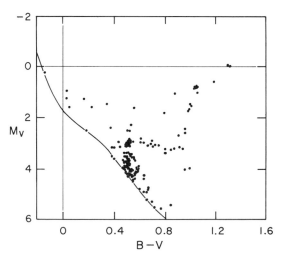

Figure 10.7. NGC 188. This is considered the oldest known open cluster. Near the point at which stars have begun to veer off the main sequence are several W Ursae Majoris eclipsing stars (EQ, ER, ES, EP Cephei). This much-studied cluster is suspected to be richer in metallic atoms than the sun, though not as rich as the Hyades. (Photograph by Harvard Observatory.)

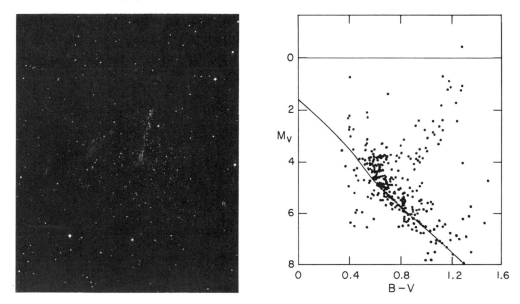

the nearby Hyades. The area covered by a cluster will always include some interlopers, which can be weeded out to some extent by a study of motions. The more distant a cluster, the more uncertain does this criterion become, both because the apparent motion of the cluster itself will be small and because its members are faint and stand out less clearly from field stars than those of a nearby cluster.

Table 10.1. Positions of seven old open clusters. Those bright enough to be seen with the naked eye or with binoculars are marked with asterisks.

Name	Position			
	h	m	°	'
NGC 188	0	35.1	+84	47
NGC 752	1	51.8	+37	11
*NGC 2281	6	42.0	+41	10
NGC 2477	7	48.7	−38	17
*NGC 2682 (M67)	8	45.0	+12	4
NGC 6940	20	30.4	+28	58
NGC 7789	23	52.0	+56	10

Figure 10.8. Comparison of the courses of the color–magnitude arrays of the old clusters, showing the progressive shearing off of the main sequence.

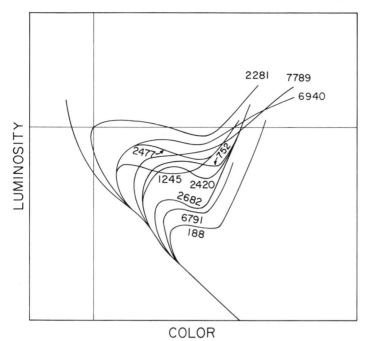

When all this is said, it still remains probable that some of the stars (known as blue stragglers) that lie above the subgiants in several of these clusters are really members. The brightest star in NGC 2682, of spectral class B8, almost certainly belongs to the cluster (fig. 10.6). How does it happen that this old group contains a relatively bright and relatively unevolved star? This problem will recur for the even more advanced globular clusters. Possibly this and similar mavericks are late-born stars, perhaps recently formed from the remains of massive, luminous stars that have run their course and returned their substance (or much of it) to the cluster substratum. Or perhaps they are double stars, which are notorious for their departures from the orthodox pattern.

Though we might have hoped to find red variable stars in these old clusters, which are relatively rich in red giants, we are again disappointed. The semiregular variable FG Vulpeculae is a possible member of NGC 6940; it is a bright giant of spectrum M5, and varies by three tenths of a magnitude (fig. 10.2). But it is exceptional.

There are four eclipsing binaries in NGC 188, the fifteenth and sixteenth magnitude EQ Cephei, ER Cephei, ES Cephei, and EP Cephei.

All are revolving nearly in contact, with orbital periods near a third of a day. They probably lie rather close to the main sequence and can therefore be only slightly evolved, reminding us of the unevolved binary TX Cancri that is a member of Praesepe. Their presence in NGC 188 could throw light on the history and development of close binary systems. The age assigned to NGC 188 depends on the conditions assumed as the basis of figure 3.4, and makes it the oldest open cluster known to us.

These old clusters tend to be fairly populous, like NGC 7789 (though NGC 752 is rather sparse). During their long lifetimes they have been continuously subject to gravitational disruption, and even the richest have probably lost a large fraction of their original population in the course of their peregrinations within the galactic disk. Clusters originally less populous or less compact have probably disintegrated altogether. The detritus of old clusters survives in the myriad field stars that are not now members of recognizable groups, though vestiges of their affiliations can be traced by noting systematic trends in stellar motions. It is not hard to believe that all field stars were once members of stellar groups.

Does the stellar field furnish us with specimens of stars that are older than the oldest known open cluster? Such stars might, for instance, be subgiants of the latest and coolest spectral classes. There are in fact no recorded subgiant K or M stars. A color–magnitude diagram of low-luminosity field stars whose motion identifies them with the galactic disk has a boundary that lies close to the relationship for NGC 188 (fig. 10.9). This suggests that the age of that cluster defines a limit for the ages of the stars in the disk population. There is no evidence for the presence in the galactic disk of stars older than the oldest known open cluster.

The observation that clusters like those shown in figures 10.1 through 10.7 grow progressively poorer in luminous (and presumably massive) stars raises anew the question of the fate of such stars. They must have perished, and in fact their remains can be observed in some of the older clusters in the shape of white dwarfs. Several white dwarfs appear in figure 2.4, which shows the established members of the Hyades. White dwarfs are known, too, in Praesepe and Coma Berenices. It has been estimated that M 67 contains nearly 200. White dwarfs may represent a terminal phase of stellar development, but their presence in the Hyades, a cluster of intermediate age, shows that they are not necessarily the oldest of the stars.

The old open clusters, even the most populous ones, are not photogenic objects like the Pleiades and the twin clusters in Perseus. They do not contain a few stars much brighter than the rest, like the clusters described in chapter 7. Their main sequences have been shorn down to a

Figure 10.9. Color and absolute magnitude for old field stars; the course of the relationship for the old clusters NGC 2682 and NGC 188 is indicated schematically. The magnitudes are given the subscript c to show that they are on a slightly different system from the B and V magnitudes used in most of the diagrams that have been represented. (From O. Eggen and A. Sandage, *Astrophysical Journal*, 136 [1962]: 745. The authors remark that the distribution "is consistent with an evolutionary picture in which stars of all ages exist up to, but not exceeding, the age of NGC 188.")

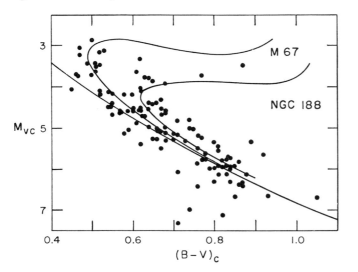

point at which many stars of similar luminosity have evolved into the sub-giant and giant stages, and accordingly they present a rather monotonous appearance. Their red giants have converged at a luminosity several hundred times that of the sun, it is true, but they are thousands of times fainter than the most luminous members of young clusters. The sun itself would cut a very poor figure in all the groups we have described; in many it would have been too faint to study.

In our survey of the stellar denizens of open clusters we have paid particular attention to variable and peculiar stars. Irregular variables, supergiants, β Canis Majoris stars, α Canum Venaticorum stars, Cepheid variables, δ Scuti stars, and semiregular red variables have successively concerned us. But several groups of variable stars, notably those most numerous in our catalogues (Mira stars and RR Lyrae stars) have been conspicuous by their absence. We have not mentioned RV Tauri or W Virginis stars and have had a mere brush with semiregular and irregular red variables. The variable stars par excellence, the novae and related U Geminorum stars, have not appeared. Neither has that rara avis, the R Coronae Borealis star.

Most of these variable stars, as we shall see, are in fact in advanced stages of stellar development. They are associated with stellar groups of a different background, members of the so-called Population II. All the groups and stars that we have hitherto described belong to Population I, which makes up the galactic disk and arms, partakes of the general rotation of the Galaxy, and (within fairly narrow limits) shares the solar composition. The representatives of Population II may be considered to form the galactic halo. They do not share the general galactic rotation, and most of them differ more or less, some very greatly, from the composition of the sun.

This description of the two stellar populations is, of course, far too simplistic. It has deliberately avoided any suggestion of age, though the subject is often approached by saying that Population I is young and Population II is old. However, after pursuing the subject of stellar development through ten chapters, we can hardly assert that a population whose members range in age from less than a million to more than ten thousand million years consists only of young stars. There can, however, be no quarrel with the statement that Population II is old, though not necessarily older than Population I.

Population I occupies the galactic disk and central plane, particularly the arm population, where star formation is still in progress. How much of the raw material that goes into making young stars is actually primitive and how much stems from the remains of stars that have run their course and perished? There must be a great deal of the latter. At least some stars are formed from material that has been recycled, perhaps several times. So a refined treatment, unlike our naive and simplistic approach, must take into account some second-generation or even third-generation stars, perhaps also stars of mixed origin, in which the products of several different recyclings are combined. Perhaps the hallmark of Population I is the continuous activity that is going on among its members.

Population II, on the other hand, seems to represent the most primitive components of the Galaxy. Although their enormous range of chemical composition raises difficult problems, it looks as though they have pursued an untroubled existence since they were first formed.

Open Clusters and the Structure of the Galaxy

The preceding chapters have drawn on the evidence provided by a number of the better known open clusters to illustrate the stages of the stellar lifetime. But clusters, moving groups, and associations are members of a greater complex, the Galaxy itself. The relationships between the stars that constitute a cluster have been used to throw light on the history of that cluster. In the same spirit we can use the relationships between the components of the Galaxy to elucidate the history of a stellar system. After examining and classifying the trees, let us take a look at the forest.

A glance at figure 1.2 reminds us that we are well within the Galaxy, though far from the middle of it. Moreover, we are very near to the galactic plane, so we view the thin central section, the habitat of young stars, nebulosity, and dust, almost on edge. The Milky Way, where we look along the galactic plane, is studded with open clusters. As we have seen, they differ greatly in age, and they also are at vastly different distances.

Over three hundred open clusters have been studied in some detail, and about a thousand have been catalogued. Actually the Galaxy must contain at least ten times this number. It is thirty kiloparsecs across, and we are about ten kiloparsecs from the center, but very few of the open clusters in our lists are more than three kiloparsecs from us. At best, our sample is representative of the local sector of our stellar system and of its nearer inhabitants.

Figures 11.1 to 11.4 illustrate the relationships of the open clusters in several sectors of the Milky Way. In figure 11.1 we stand with our backs to the galactic center and look out toward Perseus and Taurus, and from table 11.1 we note a great variety both in distance and age. The two very young clusters, α Persei and IC 348, are quite close to us, and the oldest,

Figure 11.1. Open clusters in Taurus and Perseus. NGC numbers are indicated, and the α Persei cluster, IC 348, the Pleiades, and the Hyades are marked. (Photograph by Harvard Observatory.)

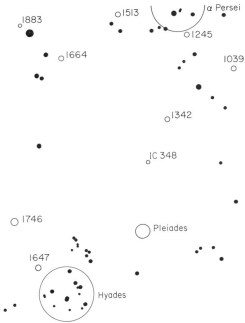

Figure 11.2. The Milky Way in Vela. A number of the clusters have not been studied in detail. NGC numbers are indicated; IC numbers are preceded by the letter I; H3 is the serial number in the Harvard catalogue of faint clusters. (Photograph by Harvard Observatory.)

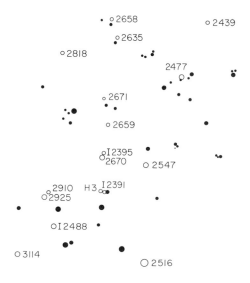

Figure 11.3. The Milky Way in Vela, Carina, Crux, and Centaurus. The region is rich in clusters, many of which have not been studied in detail. Most are young, and none are very old. *Mel* denotes the serial number in Melotte's catalogue. NGC 5139, the globular cluster ω Centauri, is conspicuous on the northern edge, at 4500 parsecs from us. Broken lines outline the Coal Sack. (Photograph by Harvard Observatory.)

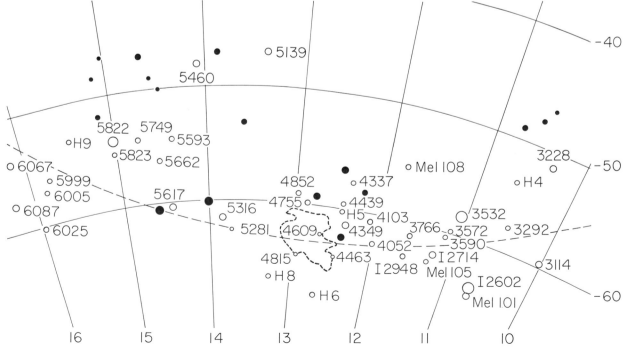

Figure 11.4.   The Milky Way in Scorpio and Sagittarius. *Tr* denotes the serial number in Trumpler's catalogue. The circled cross marks the globular cluster NGC 6121 (Messier 4), which is at a distance of 2000 parsecs. (Photograph by Harvard Observatory.)

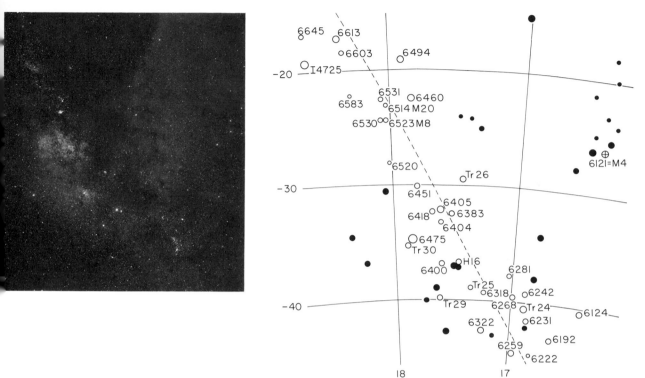

Table 11.1.   Characteristics of eight open clusters in Taurus and Perseus.

Name	Distance (parsecs)	Approximate age (millions of years)	Remarks
NGC 1039 (M34)	425	100	
NGC 1245	2300	1200	Many red giants
α Persei	170	14	
NGC 1342	550	600	
IC 348	380	<1	
Pleiades	125	70	
Hyades	42	900	Four red giants
NGC 1647	500	100	

Table 11.2.   Characteristics of nine open clusters in Vela.

Name	Distance (parsecs)	Approximate age (millions of years)	Remarks
NGC 2439	1550	8	
NGC 2477	1000	130	Many red giants
NGC 2516	330	180	Three red giants
NGC 2547	420	—	Fairly young; age undetermined
IC 2391	160	13	Brightest star is o Velorum
IC 2395	900	4	Brightest star is a binary
NGC 2670	1000	30	
NGC 2925	745	30	
NGC 3114	900	20	Contains a supergiant of class F

NGC 1245, is the most distant. In other parts of the Milky Way we shall find very young clusters at great distances, but such systems are very rare toward the galactic anticenter. The distances and ages given in this chapter are rough approximations, and investigators differ as to the actual numbers, but the tables furnish an order of magnitude.

In figure 11.2 we have turned our eyes through about 90° and are looking toward the constellation Vela in the southern hamisphere. Here again we find a variety of age and distance, but none of the clusters is very old (table 11.2). In figure 11.3, which partly overlaps figure 11.2, the brilliant star clouds of Carina display a richness and depth unrivaled in any other direction. The clusters are seen to great distances, but none has an age as large as three hundred million years, and most are much younger (table 11.3). Figure 11.4 depicts the view in the direction of the galactic center. Here the ages are still smaller, and the distances not so great as in the Carina field, though greater than most of those toward the anticenter (table 11.4).

These four samples illustrate the variety of cluster population in different sections of the Milky Way. What light can they throw on the history of star formation in our system? Young clusters will point to recent epochs of activity, old clusters to earlier proliferation. At first glance the picture seems incoherent, but a pattern emerges when clusters of different ages are examined separately.

Figure 11.5 shows the distribution of a large fraction of the well-studied open clusters, projected on the galactic plane. The four sections show clusters with ages between 10^6 and 5×10^6 years; 5×10^6 and 10^7 years;

Table 11.3.   Characteristics of twenty-two open clusters in Vela, Carina, Crux, and Centaurus.

Name	Distance (parsecs)	Approximate age (millions of years)	Remarks
NGC 2516	330	180	Two red supergiants, several binaries
NGC 2547	420	—	
IC 2391	160	13	
IC 2395	900	4	
NGC 2910	1250	130	
NGC 2925	745	30	
NGC 3114	900	20	
IC 2602	150	6	Naked-eye cluster; brightest star is θ Carinae
Melotte 101	2100	—	Red giant; age undetermined
NGC 3532	420	230	Red giants; very populous
NGC 3572	3000	1.3	Supergiant y Carinae a member?
IC 2944	2000	1	
NGC 4349	1870	130	Red giant
Hogg 15	4200	<10	Contains a Wolf-Rayet star; reddened by interstellar absorption
NGC 4609	—	—	Behind the Coal Sack, much obscured; the Coal Sack itself contains nebular variable stars such as V 755 and V 756 Centauri
NGC 4755	1000?	16	The Jewel Box; red supergiant
NGC 5316	1100	80	Red giants
NGC 5617	1100	15	Red giants?
NGC 5749	900	36	
NGC 5822	700	280	
NGC 6067	1800?	50	Red giants, possibly the Cepheid variable GU Normae
NGC 6087	850	40	Contains the Cepheid variable S Normae

10^7 and 10^8 years; and over 10^8 years. The time intervals included in the four diagrams are not the same, but are respectively five million, five million, ninety million, and over a thousand million years.

The youngest group shows clearly that these clusters are confined to relatively narrow strips, which coincide with the well-known H II regions

Table 11.4. Characteristics of eleven open clusters in Scorpio and Sagittarius.

Name	Distance (parsecs)	Approximate age (millions of years)	Remarks
NGC 6124	500	80	
NGC 6231	1700	3	Nucleus of the Scorpius I association; contains the very luminous supergiant ζ_1 Scorpii and some Wolf-Rayet stars, spectroscopic binaries
NGC 6322	1150	3.5	
NGC 6383	1200	10	Contains the eclipsing star V 701 Scorpii
NGC 6405 (M6)	500	40	Appears to contain the infrared source BM Scorpii and the very luminous early-type variable star V 862 Scorpii
NGC 6475 (M7)	240	40	Contains an unusually large number of binary stars
NGC 6494 (M23)	650	160	Two red giants
NGC 6514 (M20)	1430	—	The Trifid Nebula; very young
NGC 6523 (M8)	—	—	The Lagoon Nebula; contains a cluster of very hot young stars
NGC 6531 (M21)	1300	3	
IC 4725 (M25)	600	22	Contains the Cepheid variable U Sagittarii; the remarkable blue irregular variable V 3508 Sagittarii is nearby

(bright hydrogen nebulosities). These young systems include some bright associations and nebulous clusters such as the Lagoon, the Trifid Nebula, and NGC 6611. In the second group the roughly parallel strips can still be traced, but they are less clearly defined. In both pictures there seems to be a tendency for clusters of similar age to be bunched together, which suggests a changing timetable of star formation that sweeps forward as the galaxy rotates. The third age group shows faint vestiges of structure, but it is greatly burred. The area surveyed is smaller than in the two first pictures, in keeping with the lower luminosity of the brighter stars in the clusters. This third group comprises all the clusters that are known to contain Cepheid variables. The fourth age group shows no evidence of structure, but the area surveyed is even more restricted.

It is clear from figure 11.5 that the clusters less than ten million years old are the ones that define the observed galactic pattern. Older clusters can at best give a vague picture. The positions of all these clusters have been determined by accurate photometry, and the blurring of the picture cannot be laid to observational uncertainty. Such uncertainty, stemming from the presence of absorbing material within and in front of the cluster, is greatest for the youngest clusters, many of which are enmeshed in nebulosity, but it is these clusters that define the pattern most clearly.

Figure 11.5.   Distribution of open clusters of different ages, projected on the galactic plane. The sun is situated at the central cross; the arrow points in the direction of the galactic center. Intervals of one kiloparsec are marked off. The approximate ages are as follows: top left, 10^6 to 5×10^6 years; top right, 5×10^6 to 10^7 years; bottom left, 10^7 to 10^8 years; bottom right over 10^8 years.

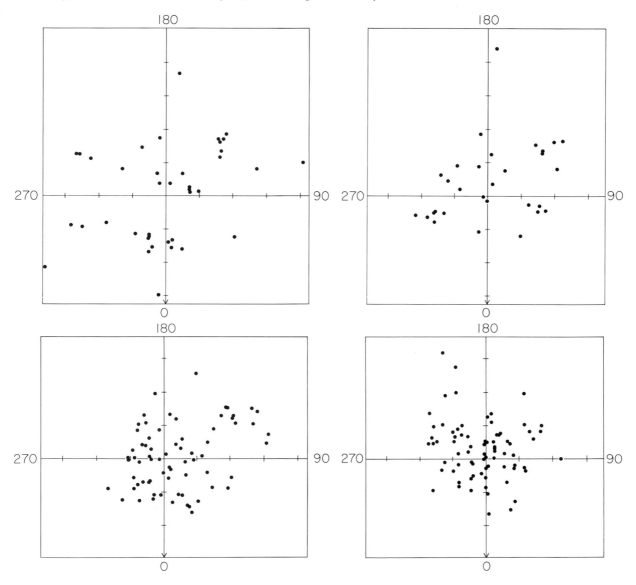

The young clusters in figure 11.5, while they show definite evidence of galactic structure, do not define the neatly convoluted spiral arms that form part of the mental picture of a galaxy like our own. Neither does the picture shown in figure 1.2, which is based on the radio observations of nebulosity that reach to much greater distances. For this reason, I have preferred to describe the structure as a swirling pattern, rather than using the more familiar term "spiral arms."

It is of interest to compare the distribution of galactic Cepheid variables, whose ages lie between 10^7 and 10^8 years (those of longest period being the youngest), with the distribution of open clusters. Figure 8.14 shows that, taken as a whole, Cepheids show little evidence of orderly arrangement, although those of the longest periods tend to do so. Most striking is the radial feature in Carina directed toward galactic longitude 290°, which represents Cepheids of long period and also of shorter period. There are few well-known clusters of the corresponding age group in the direction of Carina, though younger clusters abound there. A comparison of figures 11.5 and 8.14 reminds us that well-studied clusters do not extend over as large an area as the Cepheids do, while the radio picture of figure 1.2 embraces most of the Galaxy.

Indeed, the most compelling conclusion to be drawn from the galactic distribution of the Cepheids is that those of longest period dominate toward the galactic center, those of shortest period toward the anticenter, a tendency that may be related to the existence of a radial composition gradient in the Galaxy. The further we go from the center in the galactic plane, the poorer in metals is the galactic material. There are small differences in the composition of galactic clusters in the two directions which appear to fortify this idea. But there are very few young clusters in the direction of the galactic anticenter; this, and the absence there of long-period Cepheids, suggests that there have also been differences in the timetable of star formation in different parts of the Galaxy.

As our glance sweeps from the outer edges of the Galaxy toward the center, we see growing evidences of continued activity and current star formation. The outer boundaries are less active. Something of the same kind is suggested in the Andromeda spiral, Messier 31, where Walter Baade traced what he called faint "fossil arms" in the outer regions.

Less than a dozen Cepheids are known members of open clusters. Yet several hundred are known throughout the well-surveyed region. Were they all originally members of clusters, and if so, what has happened to the clusters? Are there unrecognized clusters or associations near to isolated long-period Cepheids?

In our survey of the open clusters, the variable stars—touchstones for the epochs and crises of the stellar lifetime—have played an important part. The irregular pre-main-sequence stars are always to be found in groups. Many of the β Canis Majoris stars populate associations and open clusters. So do variable red supergiants. Cepheids are found in open clusters, but there are far more in the general field. When we come to the irregularly variable red giants, we find that virtually none are members of open clusters; almost all are field stars. These classes of variable stars (at least those in the post-main-sequence stage) are in descending order of mass, from over thirty times that of the sun for the β Canis Majoris stars to a few times the sun's mass for the irregularly variable red giants, and are in ascending order of age, from less than a million to more than ten thousand million years.

Across the expanses of the Galaxy we have traced a moving pattern of star birth, development, and dispersal. But our story has dealt chiefly with the most massive stars. We have had little to say of those that are less massive and less luminous than the sun. In every open cluster the stage is dominated by the brightest stars. But in every known cluster the numbers of fainter stars mount up, as they do among the stars that make up the galactic field. In our own vicinity the most numerous stars are about one ten thousandth as bright as the sun, perhaps one-fortieth as massive. There is no reason to think that these figures are not typical for the Galaxy at large. Nor can we doubt that they represent the sequel to the universal scenario of star formation. These faint stars develop slowly, in harmony with their masses. In ten thousand million years even the earliest-born of them will not yet have reached the main sequence and begun to tap the great reservoir of hydrogen within them. But they, and the less massive stars that are even fainter, are not negligible members of the population. What they lack in brilliance they make up in numbers. They pass their slow uneventful lives in obscurity. But little as they contribute to the luminosity of Galaxy and clusters, they may well carry a sizeable burden of the mass.

The Globular Clusters

Open clusters and associations have their own individual characteristics and some of these, notably composition, have left their mark upon the member stars. But the open clusters have much in common, too. Their compositions differ, but not very greatly. They all lie near the plane of the Galaxy and share the galactic rotation. Seventy-nine percent of them are within 100 parsecs of the galactic plane, ninety-five percent within 200 parsecs. Only seven are more than 300 parsecs from the plane. Most of the well-studied open clusters are no more than 3000 parsecs from us, a small distance when we recall that the Galaxy has a diameter of about 30,000 parsecs and that we are about 9000 parsecs from its center. When the scene is confined to the open clusters, it remains parochial in comparison to the overall picture of the Galaxy.

Moreover, our study of open clusters has not spanned the whole range of stellar properties. It has not reached the stars of low luminosity and small mass. The variable stars, which convey so much information on the critical epochs of the stellar lifetime, are reminders that the whole story has not yet been told. The most numerous of the intrinsically variable stars, the Mira and RR Lyrae stars, have not entered the picture at all. The explosive variables, novae and U Geminorum stars, are territory still unexplored. For these, and for the stages of the stellar lifetime that they mark, we must survey the domain of the globular clusters.

About a hundred globular clusters are known in the Galaxy, and the total number is probably about twice that. The members of open clusters are counted in hundreds, but globular clusters contain hundreds of thousands or millions of stars. Consequently they are conspicuous objects, and almost a quarter of the known ones are included in Messier's catalogue.

Table 12.1 gives a list of twenty-two globular clusters, and serves to illustrate the large range of their properties. The column headed D shows the distance from us in kiloparsecs; we note that the closest globular cluster is 3 kiloparsecs away—a distance beyond which there are hardly any well-studied open clusters. NGC 7006 is 60 kiloparsecs distant, and a number of faint objects are even more remote: NGC 2419 at 69 kiloparsecs, and Palomar 4 at 125. Equally striking are the large distances from the galactic plane (column headed d): NGC 4147 is 25 kiloparsecs above it. The open clusters reveal conditions in our immediate neighborhood; with the globular clusters we range over the whole galaxy. Some of them, indeed, lie outside the usually accepted bounds of our system, and its great distance has earned for NGC 2419 the title of "intergalactic tramp."

The column headed M_{pg} gives the integrated absolute magnitude of the whole cluster in photographic light. Clearly the globular clusters are not all of the same total brightness: ω Centauri is five magnitudes brighter than NGC 6712, and a few other clusters (such as NGC 6366, absolute photographic magnitude -4.15) are even fainter than the latter. The total brightness of a cluster gives at least a rough idea of the number of stars it contains. If all the stars in ω Centauri were of the same brightness as the sun, it would contain more than a million. A star of solar luminosity would be of about the nineteenth magnitude if it were in that cluster, and such faint stars have not been studied in any cluster as yet. Therefore we do not know whether the number of faint stars mounts up in globular clusters as it is known to do for the field stars in our vicinity. However, a consideration of the dynamics of the cluster suggests that it may have a mass of about a million suns. On the basis of total magnitude we can deduce that ω Centauri contains at least two hundred and fifty times as many stars as the faintest known globular cluster. Thus the range in population is probably from several hundred thousand stars to over a million.

Only one of the clusters in table 12.1 has a galactic latitude between 90° and 270°; all the rest are in the segment 270°–0°–90°: they are concentrated in the direction of the galactic center. Figure 12.1 (bottom) shows the apparent distribution of the known globular clusters in the sky, and figure 12.2, their distribution as seen projected on the galactic plane. The confinement to the hemisphere that contains the center of our system is evident. Harlow Shapley's recognition of this fact, and his conclusion that our system is vastly larger than had previously been thought, is now ancient history. The distribution of the globular clusters is more nearly

Table 12.1. Characteristics of twenty-two globular clusters. M_{pg}: the integrated absolute photographic magnitude of the whole cluster. V: radial velocity in kilometers per second. D: distance in kiloparsecs. d: height above the galactic plane in kiloparsecs. Metal index: see text.

Name	R.A. dec. (1950) h	m	dec	Galactic Lat.	Long.	M_{pg}	Spectrum	V	D	d	Metal index
NGC 104 (47 Tucanae)	0	21.9	−72°	306°	−45°	−9.82	G3	−24	5.8	− 4.2	0.56
NGC 362	1	1.6	−71	302	−46	−7.7:	F8	+221	12.6	− 8.7	0.44
NGC 4147	12	7.6	+18	253	+77	−6.05	F2	+191	26	+25.4	0.33
NGC 5024 (M53)	13	10.5	+18	333	+80	−8.02	F4	−112	20	+19.7	0.30
NGC 5139 (ω Cen)	13	23.8	−47	309	+15	−10.40	F7	+230	5.0	+ 1.3	0.35
NGC 5272 (M3)	13	39.9	+28	42	+79	−8.49	F7	−153	13.8	+12.7	0.38
NGC 5904 (M5)	15	16.0	+2	4	+47	−7.86	F5	+49	8.3	+ 6.6	0.39
NGC 6093 (M80)	16	14.1	−22	353	+19	−6.11	F7	+18	11.0	+ 3.7	0.35:
NGC 6205 (M13)	16	39.9	+36	59	+41	−8.02	F5	−241	6.9	+ 5.2	0.34
NGC 6254 (M10)	16	54.5	−4	15	+23	−6.86	F8	+71	5.0	+ 2.7	0.36
NGC 6341 (M92)	17	15.6	+43	68	+35	−7.80	F2	−118	11.0	+ 5.7	0.25
NGC 6402 (M14)	17	35.0	−3	21	+15	−5.26	F8	−129	7.2	+ 1.8	0.42
NGC 6656 (M22)	18	33.3	−23	10	−8	−6.12	F5	−144	3.0	− 0.4	0.32
NGC 6712	18	50.3	−8	25	−4	−5.22	G4	−131	6.0	− 0.5	0.51
NGC 6723	18	56.2	−36	0	−17	−7.25	G2	− 3	10.0	− 3.0	0.51
NGC 6779 (M56)	19	14.6	+30	63	+ 8	−5.85	F5	−145	13.8	+ 1.9	0.31
NGC 6838 (M71)	19	51.5	+18	57	− 5	−5.80	G5	−80	5.5	− 0.4	0.59
NGC 6981 (M72)	20	50.7	−12	35	−33	−6.76	G0–1	−255	18.2	−11.3	0.38
NGC 7006	20	59.1	+16	64	−19	−6.85	F3–4	−348	60	−16.3	0.37
NGC 7078 (M15)	21	27.6	+11	65	−27	−8.27	F3	−107	15.1	− 6.0	0.25
NGC 7089 (M2)	21	30.9	−1	53	−36	−8.60	F3	− 5	15.8	− 8.8	0.30
NGC 7099 (M30)	21	37.5	−23	27	−47	−6.92	F3	−174	12.6	− 8.8	0.26

Figure 12.1. Apparent distribution of the open clusters (top) and globular clusters (bottom). They are marked on an equal-area projection of the sky that shows the galactic center in the middle of the diagram, marked by a cross; the central horizontal line marks the galactic plane. The sharp contrast in distribution shows that the open clusters are strongly concentrated to the plane; even those that are far from it in latitude are actually very close in distance, such as the nearby Coma Berenices, which appears at the top of the ellipse. The globular clusters, on the other hand, concentrate toward the galactic center.

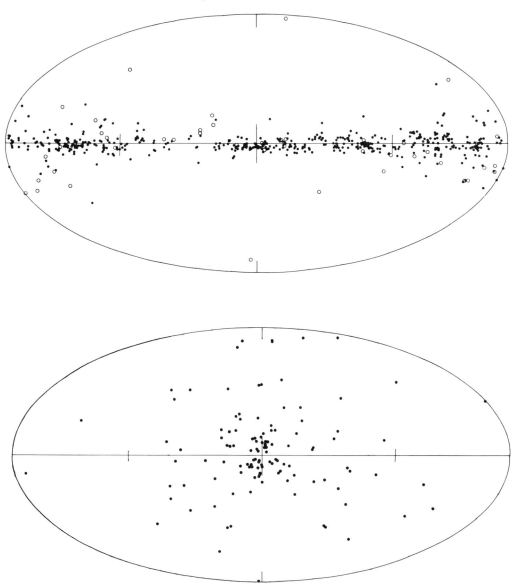

Figure 12.2. Distribution of 101 globular clusters in the galactic system, seen projected on the galactic plane. The cross indicates the galactic center, and circles of 2 kiloparsecs radius are drawn around it. A circle marks the position of the sun. Thirteen clusters lie outside the boundaries of the diagram, a distant halo or in some cases "intergalactic tramps." The clusters are distributed about the center of the system; no doubt a number remain undiscovered near the center and on the far side of it.

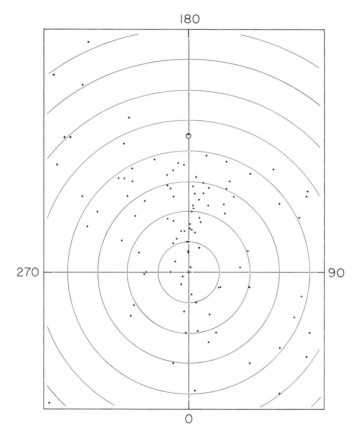

spherical than that of any other groups of celestial objects. They are the representatives of the halo par excellence. The correctness of this picture is emphasized by the similar distribution of globular clusters around other stellar systems—not only the Andromeda spiral, which is so much like our own, but also the Sombrero spiral and the great elliptical galaxy Messier 87.

The clusters, like everything else in the Galaxy, move in orbits around the galactic center under the gravitational control of the whole system. As they are confined to the galactic plane but range far above and below it, their orbits must be highly inclined to the plane. The open clusters, on the other hand, cannot be moving in highly inclined orbits, as their vertical distribution clearly shows; indeed they move in roughly circular paths about the galactic center. As the Galaxy rotates and we view the orbital motions of other objects that share the orderly rotation at different angles, their radial velocities will range through no more than ±60 kilometers a second. Anything that has a radial velocity greater than this cannot be moving in a circular orbit. A glance at the column headed V in table 12.1 shows that many globular clusters have very large velocities, and this can only mean that their orbits are not circular, but highly eccentric. The open clusters tend to move with the crowd, but the globular clusters must be passing and repassing through the plane of the Galaxy, suffering tidal disturbance as they pass through the most massive regions of the system, to which their eccentric orbits carry them. Without doubt these adventures have taken their toll, and the globular clusters we see today have been shorn of many of their former members by the gravitational effects of the central mass of the Galaxy. The least massive stars will have been the most susceptible to these outside influences. In fact, the galactic halo contains a vast number of stars that show patterns of distribution, motion, and physical characteristics that recall those of the globular clusters. These stars may well have been lost to clusters as a consequence of their repeated traumatic passages through the central regions of the system. Indeed, the present roster of globular clusters may represent the survivors of a much larger number, and may themselves have lost many of their original inhabitants.

Unlike the open clusters, globular clusters have an enormous range in chemical composition. Few are close enough for analysis of their individual stars. The brightest stars in the nearest of them are only of the tenth magnitude, whereas many of those in open clusters are naked-eye objects. But globular clusters are so populous, and their total magnitude so bright, that collective spectra have been recorded for most of them, as shown in table 12.1. The interpretation of such spectra presents special problems, because they are made up of contributions from many stars that differ in luminosity and temperature. To a knowledgeable eye, however, they can convey information that has been fortified by exact study of individual stars. The conclusion drawn from these studies is that globular clusters have a very large range in metal content from one to another, far larger than anything that has been found among open clusters or among stars that share the properties of open clusters.

The final column of table 12.1 (metal index) gives a rough empirical measure of the metal content, inferred from study of the integrated spectrum of the whole cluster and the colors of its brightest stars. Detailed spectroscopic analysis of NGC 6205 (Messier 13), metal index 0.34, gives a ratio of iron to hydrogen that is only nine percent of the solar value; for NGC 6341 (Messier 92), metal index 0.25, the ratio is less than one percent. NGC 6838 (Messier 71) has a larger metal index than NGC 6205, and NGC 6352 an even higher one (0.65); their metal content must approach the solar value. NGC 4590 (Messier 68), with a metal index 0.22, is the poorest in metals, with less than one percent of the solar value. These precise values permit the conclusion that clusters with metal index over 0.50 are relatively rich in metals, though still poorer than the sun; those with metal index less than 0.3 are relatively poor.

A few globular clusters, such as the brilliant 47 Tucanae, are relatively rich in metals; most lie in the range between one tenth and one hundredth of the solar composition, and a few have a metal content less than one hundredth of the sun's value. Thus, whereas open clusters differ in this respect by about a factor of two, globular clusters differ by two orders of magnitude.

The globular clusters in the central regions of the Galaxy tend to be the richest in metals (47 Tucanae is an exception); those that lie farther out seem progressively poorer. Thus the composition of globular clusters is in line with the growing body of evidence for the existence of a gradient of composition within the Galaxy, and may furnish a clue to the early history of our system, for they are probably as old as anything in the Galaxy. But the composition of a cluster must be related to the place where it was formed, not where it is now, and objects with highly eccentric orbits must at some time pass very close to the galactic center. If we believe that the composition of globular clusters bears the marks of their origin, we take the anthropomorphic view that these representatives of Population II are the conservatives of the stellar community, in contrast to the radical tendencies of Population I.

Globular clusters are so much more populous than open clusters that their color–magnitude diagrams seem well-defined by comparison. Even though their centers are so densely crowded with stars that accurate photometric measurements are impossible there, the number of stars that can be studied in the outskirts is great enough to define the form of the array very clearly. The number of interlopers, too, is bound to be smaller within the apparent area of a globular cluster then it is with the open clusters, most of which have larger angular diameters.

There is always a considerable number of stars of about the same magnitude and color among the brightest members of a globular cluster, and this alone suggests that such clusters are of advanced age. As we have seen, the very youngest open clusters have few stars at the upper edges of their arrays. Older clusters, even if sparsely populated, tend to have a larger number of similar, though fainter, stars among their brightest members. For this reason, the older clusters like NGC 188 are less spectacular than youngsters like the Jewel Box.

This effect is a result of the general form of the luminosity function, which expresses the relative numbers of stars of various luminosities, and thus (presumably) of various masses. The accepted picture is that the relative numbers of stars of different masses are always the same when star formation takes place. Less massive stars are formed in progressively greater numbers, with the result that the main sequence will be most populous at the lowest luminosities (at least down to absolute magnitude +16). Of course, all stars, if formed simultaneously, will not reach the main sequence at the same time, but in any one coeval group the faintest main sequence stars will be the most numerous, and fainter stars that have not matured will be even more so. As a group ages, the brighter, more massive stars are progressively shorn off from the tip of the main sequence and move toward the giant branch, subsequently disappearing from the scene. Thus the tip of the surviving main sequence grows progressively more richly populated. It is a consequence of the accepted universality of the mass frequency of stellar formation that the color–magnitude arrays of old systems are more clearly defined than those of newborn ones.

The main sequences of globular clusters are shorn down to the level of stars comparable to the sun, and here there are very many stars with the same properties. For NGC 188, the oldest open cluster we have illustrated, the highest point in the color–magnitude array, presumably representing the most evolved stage, is the bright tip of the giant branch. Every globular cluster has a well-defined giant branch, which points to a stage of development at least as advanced as for NGC 188. I am not speaking now of age, for the inferred age will depend on the composition, and the compositions of the two groups of clusters differ greatly.

Unlike the open clusters, the globular clusters do not terminate at the tip of the giant branch; the array appears to bend back again; the stars have reversed their course and are growing progressively bluer, though not changing much in visual brightness. Metal-rich clusters like 47 Tucanae only adumbrate this tendency. But the blueward extension, the "horizontal branch," grows more pronounced for clusters that are poorer in metals, and for the very poorest it sweeps back across the diagram and finally droops downward toward stars of extreme blueness and low luminosity. Well-determined color–magnitude arrays for the two globular

Figure 12.3. Color–magnitude array for NGC 5904 (Messier 5). Two type II Cepheids are shown by crosses. (See fig. 14.2 for photograph.)

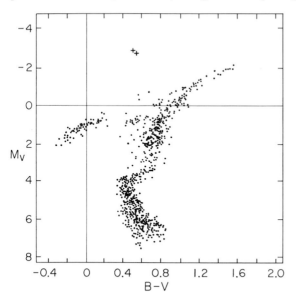

Figure 12.4. Color–magnitude array for NGC 5272 (Messier 3). The cross denotes a type II Cepheid. (See fig. 14.2 for photograph.)

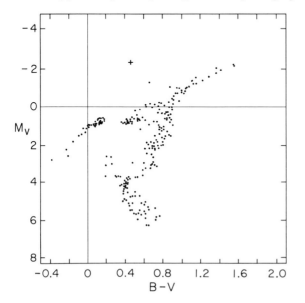

clusters Messier 5 and Messier 3 illustrate the typical globular cluster pattern (figs. 12.3, 12.4).

I imagine that the globular clusters are all of essentially the same age. When we spoke of the open clusters it was relatively easy, on account of the conspicuous progressive changes in the color–magnitude arrays, to arrange them in an order of development that corresponded at least roughly to an order of age. But this is not possible for the globular clusters. Their color–magnitude arrays differ, it is true, and the differences seem to be related to metal content, or, more generally, to chemical composition. Accordingly, I shall present the globular clusters in order of progressively smaller metal content. This choice as a criterion for grouping the globular clusters is an operational one, and does not imply a succession of ages. Among the open clusters chemical composition differs so little that it plays a small part; for globular clusters it probably takes the first place.

In all transformations within stellar communities (including both open and globular clusters), rate of development is ineluctably associated with the masses of the stars concerned. The more massive a star, the more swiftly does it develop. As the tip of the main sequence is progressively shorn down, stars of lower mass populate the main sequence. So it comes about that successive transformations of the color–magnitude array proceed ever more slowly. The brightest stars in the youngest open clusters have masses of thirty suns or more. By the time the Hyades stage is reached, about six hundred million years later, the brightest stars have two or three times the sun's mass. Those in NGC 188, nearly ten thousand million years later, are but little more massive than the sun. The brightest stars in globular clusters, on the other hand, have about the same mass as the sun. Whether or not the time scale corresponds exactly to that for the open clusters, the progress of stars within the array of a globular cluster must be extremely slow.

Metal-Rich Globular Clusters and Red Variable Stars

The globular clusters that are richest in metals are characteristically concentrated in the neighborhood of the galactic center. Many of them are faint, distant, greatly obscured, of small apparent diameter, and hence difficult to study. They can be recognized by their integrated spectra, which are of class G. These "nuclear" clusters are no doubt representatives of a large group, even more distant, more obscured, and hence apparently fainter. If their true numbers were known, the metal-rich clusters would perhaps emerge as the predominant type of galactic globular cluster. It is currently believed that there are about a hundred undiscovered globular clusters in our system, most of which must lie in the direction, and probably in the actual vicinity, of the galactic center.

Not only do these nuclear clusters appear to be small and compact, they are probably really so. Their galactic orbits must be small, so they spend their time in the hurly-burly of the galactic center, subject to the tidal forces exerted by the massive nucleus of the system.

Optical studies of these remote clusters are difficult. Progress is now being made by studies in infrared light, which is less affected by absorption than the optical region. Even more revealing are the X-ray searches: NGC 6440, NGC 6441, NGC 6712 (all metal-rich nuclear clusters) are observed to emit X-rays, and the strongest such source, NGC 6624, displays spasmodic bursts of X-rays. These remarkable observations are as yet incompletely understood; they have been associated with the compactness of nuclear clusters and the possibility that the congested centers contain "black holes."

By a happy accident one of the metal-rich clusters is also one of the nearest to us, the naked-eye object 47 Tucanae (NGC 104), which looks like a hazy fifth-magnitude star and is virtually unobscured.

Figure 13.1 is a composite color–magnitude array for three well-studied metal-rich clusters: 47 Tucanae, NGC 6838 (Messier 71), and NGC 6637 (Messier 69); the last is the richest in metals, with a metal index 0.66. The pattern of figures 12.3 and 12.4, long considered to be typical for all globular clusters, is not repeated: the horizontal branch is reduced to a mere stub. Few or no stars have moved into the blue region. There is a well-marked procession of stars from the main sequence through the subgiants to the red-giant tip, and the reddest of these are very red indeed. These optical observations are fortified by studies in the near infrared, which suggest that very cool giant tips are associated with the clusters that are richest in metals. The word "cool" is used advisedly rather than "red," since the usual optical color (deduced from a comparison of blue and visual brightness) is falsified by metallic oxide absorption, strongest for metal-rich stars. Observations in the near infrared give colors that avoid this difficulty, and show that the stars at the tip of the red giant branch are in fact cooler for metal-rich clusters, although in the optical wavelengths they are not redder. Figure 13.2 shows a color–magnitude array for 47 Tucanae, the colors being based on a comparison in visual (V) and infrared (I) wavelengths. Besides showing the extreme redness of the giant tip, this diagram reveals that the giant branch of 47 Tucanae includes a considerable number of variable stars.

The variable stars shown in figure 13.2 are of small range, but 47 Tucanae also contains three Mira variables; they all have periods near to 200 days, and are the brightest members of the cluster. These stars display the same variations of brightness and (even more significant) of spectrum that characterize Mira stars of like period in the galactic field.

Mira stars are not numerous in globular clusters. Less than two dozen have been shown to be members, and the best-attested ones are in metal-rich clusters. There are indications that the richer a cluster is in metals, the longer is the period of the associated Mira stars. But most if not all the Mira stars in globular clusters have periods less than 220 days, whereas the average for the Galaxy at large is over 300 days.

Almost 5200 galactic Mira stars are known, second only in numbers to the RR Lyrae stars, blue pulsating variables with periods usually less than a day. Stars of these two types are probably equally likely to be discovered (the chances are dependent, of course, on the structure of the discovery program, which must extend over several months for the discovery of Mira stars, while RR Lyrae stars can be found in a few weeks). They are

Figure 13.1. Composite color–magnitude array for three metal-rich globular clusters: NGC 104 (47 Tucanae), NGC 6838 (Messier 71), and NGC 6637 (Messier 69). Their distances are respectively 3.9, 3.1, and 6.3 kiloparsecs from us. The three clusters are shown below, all on the same scale. (Photographs by Harvard Observatory.)

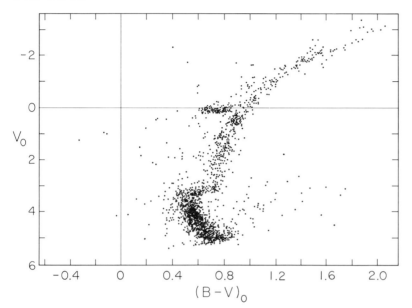

Figure 13.2. Red variables in NGC 104 (47 Tucanae). The quantity $(V - I_K)$ is an infrared color. Top: stars within 2′ of the center of NGC 104. All stars with $(V - I_K)_0 > 1.50$ are plotted but only uncrowded stars with small $(V - I_K)$. I is uncorrected for reddening. Middle: all uncrowded stars with $2′ < r < 8′$. In both plots all stars considered variable by others are plotted as V, but only new cases considered definitely variable, with a range of 0.3 magnitudes or more in V, are so indicated, though other stars are suspected to vary. Lower: the range ΔV of V magnitudes found on eight plates for stars with $(V - I_K)_0 \geq +1.00$ from the middle plot. The V magnitude is in the visual range; the I magnitude is measured in the infrared. The subscript 0 indicates that a correction for reddening has been applied. (From T. Lloyd Evans and J. W. Menzies, *I.A.U. Colloquium*, no. 21 [1972]: 157.)

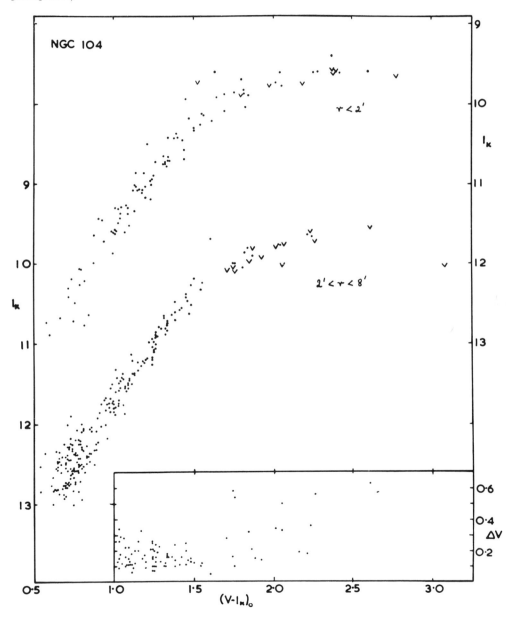

not very different in absolute visual magnitude, and as they have been found in about equal numbers in the galactic field, we may surmise that their actual numbers are comparable, at least in our vicinity. Half a dozen Mira variables are naked-eye stars at their maximum brightness, but RR Lyrae, brightest of the RR Lyrae stars, is only of the seventh magnitude. It is, therefore, remarkable that out of more than 2000 variable stars known in globular clusters, over 90 percent are RR Lyrae stars and only 1 percent are Mira stars. Their masses, so far as they are known, are not greatly different, not far from the mass of the sun. Why do Mira stars make so much poorer a showing in globular clusters than in the galactic field?

The answer probably lies in the relation of the stellar development tracks to the instability strip, for Mira stars, like Cepheids, occur within a strip that defines the surface conditions necessary to sustain variability. Perhaps the difference in temperature at the tip of the red giant branch between metal-rich and metal-poor clusters is enough to exclude Mira variation in the latter.

We need not look far for familiar specimens of the Mira type in the Galaxy. They swarm in every part of the sky, beloved of the amateur observer, who has gathered a rich harvest of information on their changing brightness, visual light curves, and periods, which range from less than 100 days to over 1000. A list of bright Mira stars is given in table 13.1. The most frequent period in the galactic field is not far from a year, but those observed in globular clusters rarely exceed 200 days. To epitomize the Mira stars in clusters we therefore should not choose Mira Ceti, the Wonder Star; its period is 332 days, and it is a double star with a very peculiar companion. A better example is X Camelopardalis, with a period of 143 days. The visual light variations of four well-known Mira stars are shown in figure 13.3, together with those of two semiregular variables of smaller range.

The spectra of Mira stars have much in common. They are dominated by molecular features, strongest when the star is faintest and coolest. Some of them show metallic oxides, others the spectra of carbon compounds; the metallic oxides show variety too, for compounds of titanium, vanadium, zirconium, and other metals appear with different relative strengths for different stars. But their most remarkable feature is the periodic emergence, as the star brightens, of bright lines of hydrogen, iron, and magnesium, which rise and fall in intensity as the star runs through its cycle of variation. The molecular bands point to a low temperature, about 3000 degrees Kelvin even at maximum when the star seems hottest, but the bright lines represent material at a much higher

Table 13.1. Characteristics of twenty-three Mira variables.

Name	Max.	Min.	Spectrum	Period (days)	DM		h	m	s	°	′
R And	6.0	14.9	S 6.6e	408.97	+37°	58	0	18	45	+38	01.4
o Cet	2.0	10.1	M52–M9e	331.65	− 3	353	2	14	18	− 3	25.1
R Tri	5.5	12.6	M4e–M8e	226.40	+33	470	2	30	59	+33	49.7
R Hor	4.7	14.3	M5e–M7e	402.67	−50	860	2	50	33	−50	17.9
R Lep	5.5	10.5	C7	432.47	−15	915	4	55	03	−14	57.4
U Ori	5.3	12.6	M6e–M8e	372.45	+20	1171a	5	49	53	+20	09.5
V Mon	6.0	13.7	M5e–M8e	334.07	− 2	1581	6	17	41	− 2	08.8
R Gem	6.0	14.0	S4–S7	369.63	+22	1577	7	01	20	+22	51.5
R Car	3.9	10.0	M4e–M8e	308.58	−62	396	9	29	43	−62	20.8
R Leo	4.4	11.3	M6e–M9e	312.57	+12	2096	9	42	11	+11	53.5
S Car	4.5	9.9	K7e–M4e	149.55	−60	2949	10	06	10	−61	03.5
SS Vir	6.0	9.6	C6	354.66	+ 1	2694	12	20	07	+ 1	19.4
R Hya	3	11	M6e–M8e	388.0	−22	3601	13	24	15	−22	45.9
R Cen	5.4	11.8	M4e–M7e	548.0	−59	5160	14	09	22	−59	26.9
R Ser	5.7	14.4	M6e–M8e	356.75	+15	2918	15	46	05	+15	26.2
RR Sco	5.0	12.4	M6e–M8e	279.74	−30	13626	16	50	15	−30	25.2
X Oph	5.9	9.2	M5e–M7e	334.22	+ 8	3780	18	33	34	+ 8	44.8
R Aql	5.7	12.0	M5e–M8e	293.0	+ 8	3970	19	01	33	+ 8	04.7
χ Cyg	3.3	14.2	S7–S10	406.84	+32	3593	19	46	43	+32	39.6
T Cep	5.4	11.0	M5e–M9e	387.79	+67	1291	21	08	13	+68	05.0
S Gru	6.0	15.0	M8e	401.54	—	—	22	19	56	−48	56.4
R Aqr	5.8	11.5	M7e	386.92	−16	6352	23	38	39	−15	50.3
R Cas	5.5	13.0	M6e–M8e	430.97	+50	4202	23	53	20	+50	49.9

temperature than this, indicating that conditions in the varying envelopes of the Mira stars must be very complex. There is little doubt that periodic disturbances are passing outward through the tenuous cool envelopes, and that the Mira stars are spilling off material with each upward surge. The coexistence of a molecular spectrum and a bright-line spectrum is the accepted earmark of a Mira star. The long-period variables in 47 Tucanae, for example, were accepted as Mira stars only when this characteristic spectrum was observed for them.

The capsule description just presented shows that the problems presented by the Mira stars are extraordinarily complex. Paul Merrill, who made a great early contribution to the study of their spectra, pointed out so many apparently contradictory details that I used to tease him by telling him that he was an obscurantist who believed that their problems would never be solved. The fact remains that up to now they have defied analysis.

Figure 13.3. Visual light curves of four Mira variables and two semiregular red variables. The Mira variables are: X Camelopardalis (period of 143 days), o Ceti, or Mira (332 days), T Camelopardalis (374 days), and S Cassiopeiae (611). The semiregular variables are: V Bootis (258 days) and R Doradus (336). All the observations were made by members of the American Association of Variable Star Observes during the interval Julian Day 2426500 to 2427900 (June 1931 to March 1934).

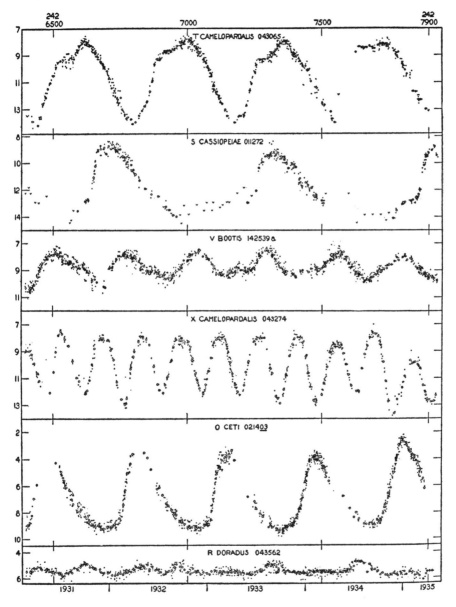

A new dimension is being added to the study of the Mira stars by observations made in the infrared. At these long wavelengths (20,000 Angstroms and more) they prove to vary through little more than a magnitude. The large changes of visual brightness are illusory, caused by the interplay of changes in energy distribution (enormous at visual wavelengths as the low temperature varies) and the absorption of light by the molecular bands. Studies of molecular spectra in the far infrared and radio regions are also helping to clarify the atmospheric structure of these huge, low-density stars, which (like the Cepheids) are probably undergoing some sort of pulsation. Moreover, the infrared and radio observations show that the Mira stars are surrounded by envelopes of circumstellar dust, probably both the product of their activity and the source of their peculiarities.

Mira stars like those in globular clusters represent the tail end of the period distribution of all known Mira stars, for only about one field Mira star in ten has a period less than 200 days. The greatest number of periods is found at about 275 days, and the average period is over 300 days. The number of field Mira stars comparable to those in globular clusters may in fact be even smaller than this comparison suggests.

To make this point it is necessary to compare the physical properties of the Mira stars in globular clusters and in the galactic field. We have seen that those in 47 Tucanae resemble field Mira stars in color and spectrum. Are they also similar in luminosity? With few exceptions (binary systems), the luminosities of Mira stars must be studied statistically by analysis of apparent motions. The results of such an analysis are shown in figure 13.4: the shorter the period, the brighter the absolute visual magnitude at maximum. Data for the Mira stars that have physical companions of known luminosity (such a X Ophiuchi, a binary with a K giant) are in reasonable harmony with the statistical results. The point that represents the stars of shortest period (average 131 days) is out of line, and so is the luminosity of X Monocerotis (155.7 days). The Mira stars, like the Cepheid variables and the RR Lyrae stars, are probably pulsating with the natural periods appropriate to their physical condition (principally their mean density).

There are in fact a number of galactic Mira stars with periods less than 150 days; they are found in Sagittarius, toward the galactic center (though not far enough away to be in the central region). These stars, many of which have the characteristic Mira spectrum, tend to have smaller ranges than the conventionally defined six magnitudes. They grade into a group of periodic and semi-periodic stars of still shorter period and smaller range, whose properties grade into those of the Mira stars. Their spectra

Figure 13.4. Period and absolute magnitude for Mira stars. Dots: absolute visual magnitudes for several groups of stars, deduced from their motions. Circles: absolute bolometric magnitudes (total radiation) for the same groups, deduced from estimates of their temperatures. The apparent decline of luminosity with period in the visual region is a result of molecular absorptions in this part of the spectrum. Conversion to total radiation shows that the Mira stars, like the Cepheids, tend to be most luminous at the longest periods.

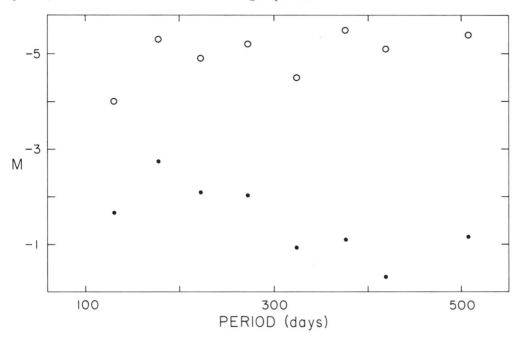

display variable molecular bands, and bright lines are observed at comparable phases of their light curves. Perhaps they should not be considered physically distinct from the Mira stars. About thirty such variable stars are known, and their luminosities make them brighter than the Mira stars: visually and photographically they are more luminous than the average Mira stars of much longer period.

But visual magnitude is a deceptive index of luminosity. The Mira stars of longest period are those of most advanced spectral type, and accordingly the coolest. The cooler a star, the more its light recedes into the unseen infrared. When the appropriate adjustments are made, all the Mira stars appear to have about the same total (or bolometric) luminosity, including the short-period, small-range stars just described. Although similar in luminosity, they are of course not similar in size or mean density. The brightness of unit surface varies as the fourth power of the temperature; since surface area varies as the square of the radius, the radii of a series of stars of equal luminosity would vary inversely as the squares of their temperatures. The temperatures are very uncertainly known, and this argument is a crude one in that it pictures the stars as glowing spheres with uniform surfaces at one given temperature; but it does suggest the likelihood that the Mira stars of longest period are the largest. That the periods are related to the mean densities, as for Cepheids, while not shown in detail, appears at least to be plausible.

The point of contact between the Mira stars in globular clusters and those in the galactic field is provided by the three well-observed Mira stars in 47 Tucanae (fig. 13.1). We may conclude that they resemble the galactic field stars in luminosity, as they are known to do in spectrum. Thus, we link the short-period Mira stars of the galactic field with representatives of one of the globular clusters that is richest in metals, bearing in mind that within the globular clusters, the greater the metal richness, the longer the periods of the Mira stars.

The tail end of the period distribution of all Mira stars, and its comparability with members of metal-rich globular clusters, gives but a tenuous grip on the part played by the Mira stage in stellar history. But it does furnish a *terminus ad quem*. The Mira stars (save for a possible discontinuity introduced by overtone pulsations) show a continuous distribution in period, spectrum, and visual luminosity. This is not all. Their distribution within the galactic system and the character of their motions show continuous progressions with period. Those of shortest period are least concentrated, and those of longest period most concentrated, to the plane of the Galaxy. The globular clusters are even less concentrated to the galactic plane than Mira variables of shortest period, whereas the open clusters are far more concentrated than those of longest period. Similar tendencies are shown by their systematic motions. Thus, Mira variables form a sort of bridge between the two types of cluster. Those stars with shortest periods verge toward the halo; those with longer periods grade continuously into the disk.

The short-period Mira stars are slightly metal deficient (with about a quarter of the sun's metal content), and recall the similar composition of globular clusters like 47 Tucanae. Their masses are probably about like the sun's, and their ages, like those of the clusters, of the order of 10 billion years. Those of longer period are of more nearly solar composition. They are probably rather more massive, and (to judge from their distribution and motions) are progressively younger, the longer the period—but probably not by very much. The subsequent adventures of a long-period variable can perhaps be inferred by looking for other objects of similar distribution and motion.

Finally, it should be mentioned that the Mira stars of the Galaxy show

Figure 13.5. Frequency of period for galactic Mira stars in different longitudes. Left: 400 square degrees centered on the galactic center. Middle: 1600 square degrees centered on the galactic anticenter. Right: all others. The concentration of shorter periods toward the galactic center is evident.

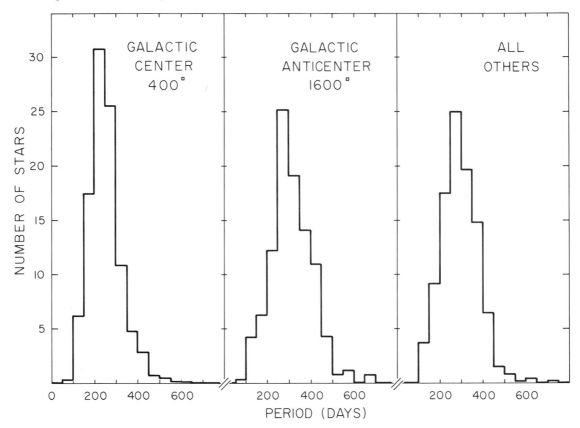

a strong concentration toward the galactic center for periods less than 250 days. This is more than 50 days less than the most frequent period in the system as a whole and is consistent with the possibility that most of these stars had their origin in metal-rich globular clusters, also concentrated to the center of the system. However, the Mira stars in other parts of the Galaxy have on the whole longer periods and greater metallicity (fig. 13.5). Are they relics of clusters of still greater metallicity than the nuclear globular clusters? If they are more massive than those of shorter period, their relatively large numbers call for explanation, both because more massive stars are less frequent in the Galaxy and because (other things being equal) they develop more swiftly.

The relation of Mira variables to the second giant branch of metal-rich globular clusters leaves many questions unanswered. Why are there none in open clusters, even in the oldest? And why are they so numerous in the galactic field? The Mira variables are in some ways the least understood of variable stars. The intricate variations of their spectra still defy analysis. However, modern studies in the infrared and radio regions, which are revealing circumstellar envelopes for many of them, suggest that a physical picture of these, the coolest and more gigantic of variable stars, is within reach.

Besides the Mira variables, these clusters contain a number of irregular and semiregular variables of small range. In fact, all stars that are as red as the extreme tip of the giant branch may well be variable. The situation shown in figure 13.2 is probably typical. For more irregular and semiregular variables have been recorded in globular than in open clusters, and not a single Mira star is a known member of an open cluster. Even when we recall that globular clusters are at least a thousand times more populous than open clusters, the contrast is striking.

Viewed on the background of the globular clusters, the Mira stars can be regarded as marking yet another critical epoch in the stellar lifetime. They are in transition as they pass into a zone of instability. They are characteristic members of clusters which show a mere stub of the horizontal branch.

When considering the sequence of development of open clusters, we were able to relate it to a relatively simple series of tracks calculated for a single choice of chemical composition. But such calculations for globular clusters are complicated by the huge range of metal content that must be taken into account. Instead of relatively simple tracks, we are confronted by interweaving patterns, and it becomes more and more difficult to match theory and observation.

Instead of making an orderly blueward progression from the tip of the giant branch along the horizontal branch, the star is expected to leave the horizontal branch at some point whose position is dictated by composition, mass, and structure. It turns back and approaches the red tip for a second time (the "asymptotic giant branch"), and the point at which it turns back determines the extent of the horizontal branch—for the metal-rich clusters, a mere stub. The Mira stars are associated with this second redward excursion.

Globular Clusters and Short-Period Variables

The form of the color–magnitude array of a globular cluster is evidently related to metallicity. The metal-rich clusters have only a vestigial horizontal branch, but for clusters that are poorer in metals the horizontal branch extends from red to blue stars, and for those of smallest metallicity its blue end grows progressively more populous. Within the color–magnitude array there are two instability strips: one at the extreme end of the giant branch and another that cuts a swath across the relatively blue stars. The former is the domain of the Mira stars. The latter crosses the horizontal branch and defines a small range of colors within which virtually all the stars are variable in brightness. We are reminded of the Cepheid instability strip that lies in the middle of the Hertzsprung gap. Here we find the RR Lyrae stars. They are sometimes called "cluster-type variables," but I avoid this misnomer because they are by no means confined to clusters, nor are they the only types of variable stars found there.

In his classical studies of variable stars in globular clusters, Solon I. Bailey concentrated, by a happy instinct, on Messier 3 and ω Centauri. These two clusters contain large numbers of RR Lyrae stars, blue pulsating variables with periods usually less than a day, which resemble Cepheids in many ways; they are named for their brightest representative, a galactic field star. More than 90 percent of the known variables in globular clusters are RR Lyrae stars, but they have not been found in all clusters. In fact, as Helen Sawyer Hogg has pithily put it, "the most frequent number is zero." They are far more numerous in the galactic field than in globular clusters, over five thousand having been catalogued.

Table 14.1 summarizes the RR Lyrae population of twenty-two well-studied globular clusters. The metal-rich clusters are evidently the

CHAPTER 14

Table 14.1. Characteristics of twenty-two globular clusters.

Cluster (NGC)	Metal index	Integrated spectrum	Type	RR Lyrae stars	
				Observed	Adjusted
6637 (M69)	0.66	G5	—	0	0
6838 (M71)	0.59	G2	—	0	0
104 (47 Tucanae)	0.56	G3	—	2	0.5
6171 (M107)	0.50	G2	I	22	69
362	0.44	F8	I	12	17
6362	0.43	—	I	23	78
5904 (M5)	0.39	F5	I	92	46
6981 (M72)	0.38	G2	I	33	69
5272 (M3)	0.38	F7	I	200	97
5139 (ω Cen)	0.35	F7	II	137	11
6093 (M80)		F7	—	?	?
6752	0.35	F6	—	?	?
6205 (M13)	0.34	F5	—	2	1
6397	0.33	F5	—	0?	0?
6656 (M22)	0.32	F5	II	17	9
4833	0.31	—	II	11	11
7089 (M2)	0.30	F3	II	17	5
5024	0.30	F4	II	33	12
5897	0.26	—	?	5	7
7099 (M30)	0.26	F3	II?	3?	5?
7078 (M15)	0.25	F3	II	78	23
6341 (M92)	0.25	F2	II	13	8

poorest in RR Lyrae stars. None have been found in the two richest (Messier 69 and Messier 71), and 47 Tucanae, one of the most populous, has but two. A third lies nearby; it is of the same brightness as the other two, and is accordingly at the same distance from us, but its measured motion seems to exclude it from actual membership. It is tempting to think that this star may be a lost member of the cluster. Many years ago Harlow Shapley touched upon the same idea: "I used to think," he remarked to me ruefully, "that they were trailing clouds of glory," that the RR Lyrae stars of the galactic field were stars that had been lost to globular clusters. He may have been right after all. Globular clusters are thought to have lost a great many of the stars that they originally contained.

In assessing the richness of a cluster in RR Lyrae stars, however, its total population must be taken into account, and the populations of globular clusters differ by a factor of at least a thousand. The bright 47 Tucanae is exceptionally rich in stars, with probably over a million members, and two RR Lyrae stars are a very small sample of them. There are 137 known RR Lyrae stars in ω Centauri and only 10 in NGC 5053, but in proportion to total number of members, NGC 5053 is three to four times richer in RR Lyrae stars than its more densely populated fellow. In comparing numbers of RR Lyrae stars we have therefore adjusted them arbitrarily to a standard of total visual cluster luminosity ($M_v = -7.5$), which gives a rough measure, at least, of the true proportions. The adjusted numbers are given in the last column of table 14.1.

Figures 14.1 through 14.5 show five color–magnitude diagrams that have been compounded from the data for nineteen of the globular clusters in table 14.1, which also contains the three metal-rich clusters described in chapter 13. The table is divided into six sections of decreasing metallicity index. The type of cluster (I or II) refers to the distribution of the periods of the RR Lyrae stars, as explained below.

In figure 14.1, which represents the three clusters with metal index between 0.50 and 0.43, the red end of the horizontal branch is well populated with stars, but the stub extends farther to the blue. Then comes the gap that is populated by the RR Lyrae stars, and there is a mere scattering of still bluer stars. Unlike the metal-rich clusters, those shown in figure 14.1 contain respectable numbers of RR Lyrae stars.

Figures 14.1 through 14.5, which show clusters of progressively smaller metal content, display a concomitant change in the distribution of stars along the horizontal branch. The red side of the RR Lyrae gap grows less populous and the blue side more populous with decreasing metallicity. Most of these clusters contain RR Lyrae stars, but in smaller numbers where the metallicity is low.

In looking at these diagrams, we must remember the difficulties that attend the detailed study of globular clusters. Because of the crowding of stars in the center, the observations necessarily represent the outer, sparser regions. Then, too, the total number of stars is so enormous that selections must be made, and a larger proportion of the brighter stars will be represented in the diagrams. Moreover, there is an inevitable cutoff for the faintest stars that depends on the distance of the cluster from us as well as on the instrument available to study it, so that undue significance should not be attached to the extreme blue "tail." Thus, the diagrams give an idea of the *form* of the color–magnitude array but not (unless special precautions are taken) of the relative numbers of stars in different parts of it. That the number of stars in all these clusters is increasing down to the observed limit must be considered certain, but an accurate determination of the luminosity function or functions presents special problems.

Figure 14.1.   Composite color–magnitude arrays for the three globular clusters NGC 6362, 362, and 6171 (Messier 107), respectively 5.5, 7.6, and 5.2 kiloparsecs from us. They are moderately rich in metals, but poorer than those described in chapter 13. Below: photographs of NGC 6362 and 362, on the same scale. (Photographs by Harvard Observatory.)

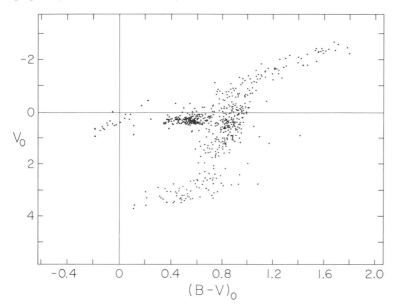

Figure 14.2. Composite color–magnitude array for the three globular clusters NGC 5272 (Messier 3), 6981 (Messier 72), and 5904 (Messier 5), respectively 8.8, 15.4, and 6.7 kiloparsecs from us. They are poorer in metals than those of figure 14.1. Below: photographs of NGC 5272 and 5904, on the same scale. That of NGC 5272 is on a red-sensitive plate. (Photographs by Harvard Observatory.)

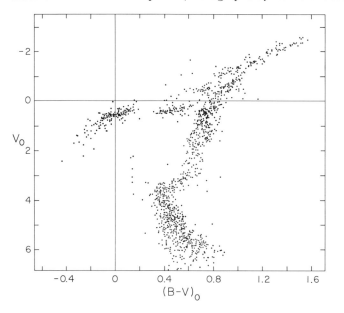

Figure 14.3. Composite color–magnitude array for the four globular clusters NGC 6205 (Messier 13), 6752, 6093 (Messier 80), and 5139 (ω Centauri). Their distances are respectively 6.3, 4.0, 9.7, and 4.9 kiloparsecs from us. They are poorer in metals than those in the two preceding figures. Below: photographs of NGC 6205, 6093, and 5139, all on the same scale. (Photographs by Harvard Observatory.)

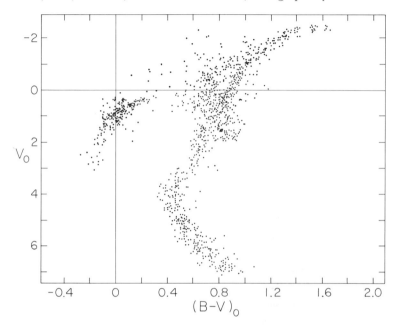

Figure 14.4. Composite color–magnitude array for the five globular clusters NGC 5024 (Messier 53), 7089 (Messier 2), 4833, 6656 (Messier 22), and 6397. They are poorer in metals than those in the three preceding figures. Below: photographs of NGC 4833, 6656, and 6397, all on the same scale. Their distances are respectively 5.1, 2.7, and 2.0 kiloparsecs. (Photographs by Harvard Observatory.)

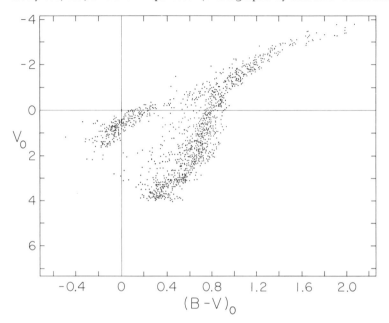

Figure 14.5. Composite color–magnitude array for the four globular clusters NGC 6341 (Messier 92), 7078 (Messier 15), 7099 (Messier 30), and 5897. Their distances are respectively 7.5, 9.8, 7.4, and 10.8 kiloparsecs. These are the poorest in metals of the clusters illustrated. Below: photographs of the clusters, all on the same scale. (Photographs by Harvard Observatory).

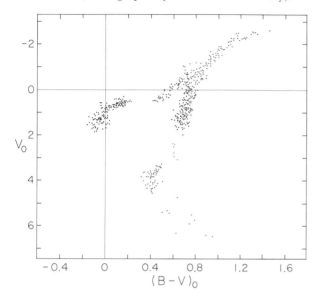

The numbers of variable stars, as given in table 14.1, are uncertain, too. In many cases the numbers of known RR Lyrae stars with determined periods is smaller than the number of rapidly varying stars that have been recorded, many of which must be RR Lyrae stars. Much work still remains to be done on these stars, and also on the discovery of such stars in clusters that have been incompletely studied. Therefore, the actual numbers, and also the normalized numbers of RR Lyrae stars, must be regarded as lower limits. The general trends, however, are so well-marked as to carry conviction. Each cluster, of course, has its own individuality. Although all are probably of about the same age, this is not necessarily true in detail, but to assign exact ages, or even to arrange clusters in order of age, is not now possible. We have merely arranged them in order of metallicity.

The RR Lyrae stars are pulsating variables, almost all with periods less than a day. Typical light curves show two distinct patterns (fig. 14.6). Some stars show a large and rapid rise in brightness, often preceded by a slight "hitch" and followed by a slower decline. Others show a more or less symmetrical change in brightness, nearly a sine curve, of much smaller magnitude range. The asymmetrical curves of large range go with the longest periods, the small-range symmetrical curves with the shortest periods, and in any one well-populated cluster there is a more or less pronounced range of period in between, within which no stars are found. It is generally agreed that the longer periods correspond to a fundamental vibration, the shorter periods to the first overtone vibration, and the relation between the periods is what would be expected on the basis of theory.*

---

* Overtones are familiar to anyone who has played on a wind instrument. The pitch of the note is determined by the length of the vibrating column of air, which in turn is governed by the player's manipulation of the stops. The length of the column determines the wavelength of the sound wave produced when the instrument is gently blown; the wavelength is inversely proportional to the frequency of the note emitted, so that a long column gives out a low note, a short column a higher one. The density of a vibrating star is analogous to the length of the air column: a star of low density has a long period of vibration; one of high density vibrates (or pulsates) with a shorter period. But let the player blow more vigorously, and the pipe will sound not the original note but the octave or first harmonic: the column has been broken up into two, and the note that emerges is the same as would have been produced by gently blowing on a pipe of half the length. Originally the note emitted was the *fundamental*: overblowing has produced an *overtone*. In a star the overtone is not of exactly twice the frequency (and therefore half the period) of the fundamental: the star is not of uniform density like the column of air that permits a pipe to "discourse most eloquent music." Both the pipe and the star are also capable of higher overtones, produced when the column of air breaks up into three, four, or more segments. In fact all vibrations, both of pipes and stars, involve different combinations of fundamental and overtones in various proportions.

This division into two period groups with differing light curves is evident for all globular clusters that contain more than a few RR Lyrae stars. But the actual periods are not the same for all clusters. They fall, in fact, into two well-differentiated groups, illustrated in figure 14.7 for NGC 5272 and ω Centauri. In the former cluster the transition from fundamental to first overtone is seen at about half a day; in the latter, at about six tenths of a day. The clusters with the lower transition period are designated type I, the others, type II. These types are given in table 14.1, for the clusters represented in figures 14.1 to 14.5. The RR Lyrae stars of the galactic field show a similar dichotomy of periods, but it is less well marked than for either of the cluster types; presumably it represents a mixture of the two.

That there is a close relation between the metallicity of the clusters and the features presented by their RR Lyrae stars is evident in table 14.1. It is more succinctly presented in tables 14.2 and 14.3, which show average values for the five groups. In type I clusters, which have the higher metallicity, the RR Lyrae stars are more frequent and their transition period is shorter. They are also found to be somewhat less luminous.

It would carry us too far afield to attempt to give an account, even a brief one, of the efforts that have been made to interpret this information. The situation is complicated by the number of parameters. Composition has innumerable possibilities. Is the metallicity, for example, uniquely related to the vital helium content? To what extent have the brightest cluster stars lost mass during their red-giant phase or phases? Even if all globular clusters were of the same age, would stars in clusters that differ in chemical composition have exactly the same physical properties (luminosity, mass, structure) when they arrived at analogous points on their tracks of development? Probably not. We have mentioned that the RR Lyrae stars in clusters of types I and II differ in transition period, luminosity, and real frequency. Probably they also differ in mass. This is the simplest conclusion, because the fact that they are pulsating places them at an identifiable stage of development associated with known physical conditions.

The RR Lyrae stars, like the Cepheids, populate an instability strip— the same one, in fact, that is the domain of the Cepheids and the δ Scuti stars. The tracks of development may not move the star far enough to the blue to cross the instability strip at all (as in the metal-rich clusters), either because there has not been time or, more probably, because the extent of the bend is limited by the star's inner economy. Or, like the Cepheids, they may cross and recross the strip, moving alternately blueward and redward through it.

Figure 14.6. Light curves of some of the RR Lyrae stars in the globular clusters NGC 5139 (ω Centauri). The shorter periods show the symmetrical form characteristic of the first overtone pulsation; at about period 0.474 days we note the transition to the asymmetrical light curve associated with the fundamental. (From a study by W. C. T. Martin, *Bulletin of the Astronomical Institutes of the Netherlands*, 18.)

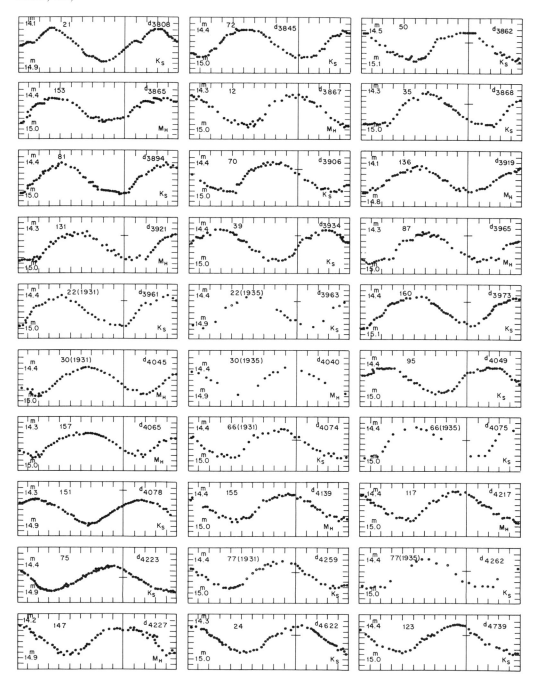

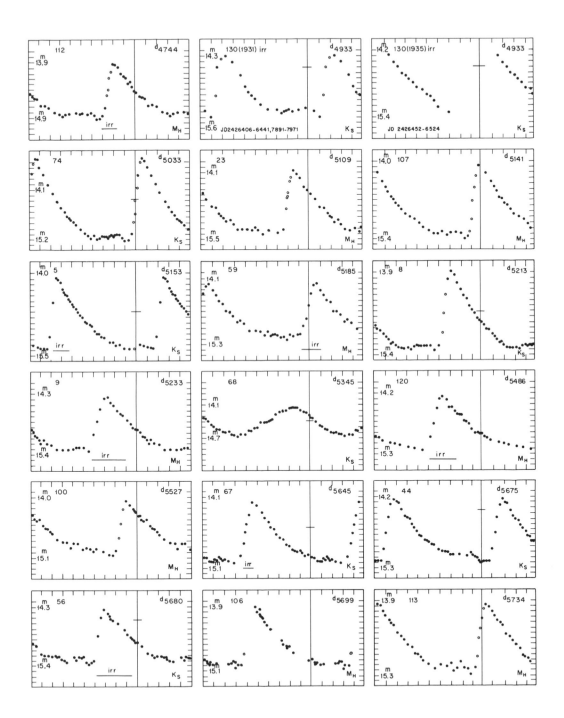

Figure 14.7. Frequency of periods of RR Lyrae stars in NGC 5139 (ω Centauri) and Messier 3. The distribution for the former is characteristic of type II, the latter of type I.

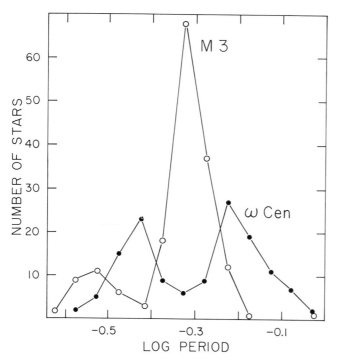

Here is a possible basis for understanding three classes of globular clusters: those without RR Lyrae stars, those with a short transitional period, and those with a long one. The tracks for the first group may never reach the instability strip. In the second group the stars on the red-to-blue track that are pulsating with the fundamental period are reluctant to break the

Table 14.2. Metallicity and properties of globular clusters and number of RR Lyrae stars.

Average metallicity index	Cluster type	Normalized number of RR Lyrae stars
0.46	I	55
0.38	I	71
0.35	II	3
0.31	II	7
0.26	II	11

Table 14.3. Metallicity of globular clusters and properties of their RR Lyrae stars.

Cluster type	Average metallicity index	Normalized number of RR Lyrae stars	Luminosity of RR Lyrae stars	Transition period (days)	Changes of period
I	0.42	63	fainter	0.464	increase, decrease
II	0.29	7	brighter	0.57	increase

habit when they arrive at the point characteristic of the first overtone, and overshoot the transitional period slightly. On the blue-to-red track (the third group) the stars pulsating in the first overtone have a similar reluctance to vibrate with the fundamental period, and again overshoot the critical transitional period. Because period is dictated by density and a given star increases in density as it grows bluer, a star on the red-to-blue track will tend to have a diminishing period, one on the blue-to-red track an increasing one.

The last column in table 14.3 relates to observed changes of period, which are commonly found in RR Lyrae stars both in and out of clusters. These changes of period are problematical, for they tend to be abrupt and therefore cast doubt on an interpretation based on steady, systematic, physical change. However, the tendency of RR Lyrae stars in clusters of type II to increase in period would be consistent with a position on the redward track. The mixture of increasing and decreasing periods in clusters of type I is less simple. It could, optimistically, be related to a doubling back of the track upon itself. But this stretches the interpretation very thin, and it is not sure that the observed changes of period are in fact of the size that the theory requires.

Irrespective of these complexities, the presence of RR Lyrae stars in globular clusters and their occurrence in a small and well-marked area of the color–magnitude array must be seen as marking a definite stage in the stellar lifetime. The low luminosity at which stars in globular clusters leave the main sequence and the large number of subgiants both point to considerable antiquity. They also assure us that the masses of stars at this critical point are rather less than that of the sun. The RR Lyrae stage comes after the first red-giant stage and probably before the Mira stage; perhaps some very metal-rich clusters never develop RR Lyrae stars.

There are many globular clusters—indeed the majority—in which no RR Lyrae stars have been found at all. There can be a number of reasons for this. The clusters may be metal-rich; they may, like NGC 6397, have a small population; the known variable stars may have been incompletely studied, as in IC 4499, where over a hundred variable stars are known but no periods have yet been found; or the clusters may never have been searched for variable stars on account of faintness and inaccessibility, though this group is happily dwindling.

The RR Lyrae stars of the galactic field are very numerous indeed; over five thousand have been catalogued. The ratio of known cluster specimens to known field specimens is 0.22, whereas the corresponding ratio for the Mira stars is only 0.05—despite the fact that these two types of stars are equally easy to discover and probably differ little in luminosity, mass, and only slightly in age. Unlike the Mira stars, which span the gamut between disk and halo in distribution and motion, the RR Lyrae stars belong predominantly to the halo. This allies them more closely to the globular clusters, as does their general tendency to be poorer in metals than the sun. There are, however, some RR Lyrae stars in the disk population, and they tend to have shorter periods and higher metal abundance than those of the halo. These might have originated in groups comparable to the old open clusters, as we have surmised the Mira stars of longer periods to have done. The majority of Mira stars in the field belong to this category, whereas only a very small minority of the RR Lyrae stars, those of shortest period, do so.

No RR Lyrae stars have been found in open clusters, and the existence of a true horizontal branch for such clusters remains to be demonstrated. There are a few blue main-sequence stars or near main-sequence stars in some old open clusters, as we saw in chapter 10, well above the turnoff point, and they present problems in the understanding of stellar development. Some globular clusters (ω Centauri, for example) have similarly situated members, known as "blue stragglers," that are uncommon but inescapable. Are they late-born stars, or are they double or multiple and thus subject to the vagaries induced by close companions? This question is still unanswered.

It is tempting to think that Shapley was right in thinking that the globular clusters may be trailing clouds of glory. Most of the 5100 known galactic RR Lyrae stars may once have been members of globular clusters. The parent clusters are not necessarily observable or even surviving. Nor were the present RR Lyrae stars necessarily pulsating when they were lost to the clusters, so that a direct comparison of the 1200 known in clusters with the 5100 in the field is not of obvious significance. The active lifetime of an RR Lyrae star may not be, probably is not, a unique quantity. The range of composition of RR Lyrae stars is comparable to that of the globular clusters themselves.

Besides the RR Lyrae stars, globular clusters contain a number of variable stars whose periods fall in the same range as those of Cepheids in the galactic field and in open clusters. The light curves of some of these stars are shown in figure 14.8. Their presence in globular clusters shows that they must be both fainter and older than classical Cepheids. They are represented in the galactic field by a wholly comparable group of variables, comprising the so-called W Virginis stars, known as type II Cepheids in recognition of their association with Population II (Cepheids are Population I stars), and RV Tauri stars. Light curves of eight of them are shown in figure 14.9.

Classical Cepheids show a well-marked relation between period and form of light curve. The light curves of W Virginis stars, on the other hand, show a less definite progression of light curve with period, and at any one period the form of variation is quite different. Most Cepheids have fairly constant periods, though for the longest periods there is a tendency toward systematic, gradual change of period; most of them repeat their light variations with great fidelity from cycle to cycle. The periods of W Virginis stars tend to vary erratically, and so do the forms of their light curves, which grade into the truly erratic behavior of the semiregular RV Tauri stars. The latter resemble, and tend to grade into, the semiregular variables that were mentioned as possible shorter-period analogues of the Mira stars. The spectra of W Virginia stars vary with the cycle, changing in spectral type and radial velocity in a way that recalls the Cepheids and points to pulsation of some sort. But they also show emission lines as the brightness rises, much as the Mira stars do, and there are discontinuities in the run of radial velocities which are best understood in terms of a surge of material that is ejected in the course of each cycle. Such bright lines and discontinuities of radial velocity have been observed for RR Lyrae stars, too.

Moreover, the luminosities of W Virginis stars are much lower than those of classical Cepheids of similar period. Like the Cepheids, they display a period–luminosity relation. The two relations are parallel but by no means coincident. This information comes not only from the well-determined luminosities of the W Virginis stars in globular clusters but also from studies of these stars in stellar systems other than our own. The fourteen in the Large Magellanic Cloud have a period–luminosity curve two magnitudes fainter than that for Cepheids. Data from Messier 31, the spiral galaxy in Andromeda, support this conclusion.

The W Virginis stars differ from classical Cepheids in composition, too. Cepheids evidently have the metal-rich composition characteristic of relatively young stars such as those in open clusters. Some W Virginis stars are conspicuously poor in metals—for example, AC Herculis. For the field W Virginis stars the data conspire to suggest that we are dealing with

Figure 14.8.   Light curves of six type II Cepheids in the globular cluster ω Centauri (NGC 5139). Photographic magnitudes are to the left; phases in tenths of the period are marked. In each diagram the number on the upper left is the serial number of the variable in the cluster; on the upper right is the period in days. (From the work of W. C. T. Martin.)

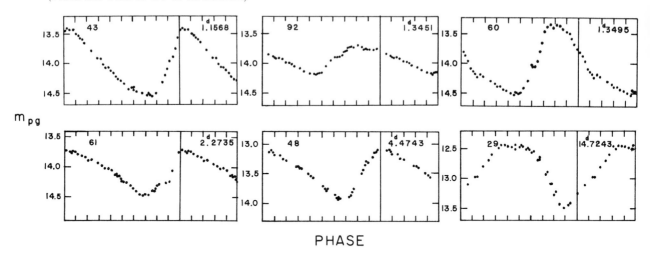

Figure 14.9.   Light curves of eight galactic type II Cepheids. Comparison with figure 8.13 shows that the forms of the light curves, and their relation to period, are quite different from those of the classical Cepheids. (1) BL Herculis (period 1.307 days), (2) VZ Aquilae (1.668 days), (3) κ Pavonis (9.064), (4) AP Herculis (10.408), (5) CS Cassiopeiae (14.733), (6) W Virginis (17.277), (7) RX Librae (24.932), (8) TW Capricorni (28.558).

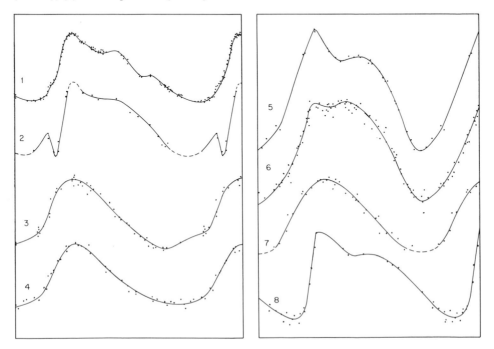

a group of old, metal-poor stars of relatively low luminosity, with masses not greater than that of the sun, whereas the Cepheids are young and rich in metals and have masses from three to nine times the sun's.

The distribution of these two classes of pulsating variable stars is as distinctive as their behavior. Cepheids are confined to the galactic plane. Only a few are as much as 250 parsecs from it. Those of the very longest periods favor the "spiral structure" outlined by the youngest clusters and the bright hydrogen nebulosities. Only an optimistic eye, however, can discern any structure in the distribution of the Cepheids at large, nor (considering the ages of the majority) would this be expected. By contrast, W Virginis stars are found at great distances from the galactic plane, up to 4000 parsecs. Their distribution shows no hint of structure. Moreover, they are clearly concentrated toward the galactic center, which strongly suggests that they are associated with the halo population and is in harmony with their presence in globular clusters. About three dozen W Virginis and RV Tauri stars are known in globular clusters, and perhaps 200 in the galactic field. Thus, the ratio of field W Virginis stars to cluster representatives is about the same as for the RR Lyrae stars. Mira stars are relatively far less frequent in globular clusters, as we noted earlier. For what it is worth, this tenuous piece of information suggests that the W Virginis stars are more closely associated in the pattern of the stellar lifetime with the RR Lyrae stars than with the Mira variables.

The periods of the Cepheid variables proper display a continuous distribution, though it is true that the *frequency* of their periods differs from one stellar system to another and even in various parts of the same system. For example, there is a difference between the distributions of period in the Galaxy and in the two Magellanic Clouds, and a definite difference of period between the center and edge of the Andromeda galaxy. But the periods of the W Virginis stars, both in the galactic field and in the clusters, fall into two groups that do not overlap at all (fig. 14.10). There is a maximum of frequency at about 2 days, a gap from 6 to 8 days, and another maximum around 16 days. Possibly the short-period group represents the tail end of the RR Lyrae distribution: the luminosities are comparable. A marked period–luminosity relation begins after the gap at a period of about 7 days; it is roughly parallel to that for Cepheids but is two magnitudes fainter. Both period groups are represented in clusters of types I and II. The long-period group preponderates slightly for type I, the short-period group for type II, but the numbers are too small to be very significant. The period–luminosity relation is shown in figure 14.11.

There is little doubt that the W Virginis stars bear the same sort of relationship to the red giants of the globular clusters that the Cepheids bear to the tip of the giant branch for the younger stars. As we saw in chapter

Figure 14.10. Frequency of period for W Virginis stars, or type II Cepheids. Above: type II Cepheids in globular clusters; below: in the galactic field. Note the separation into two groups and the general similarity in the distribution for cluster and field stars.

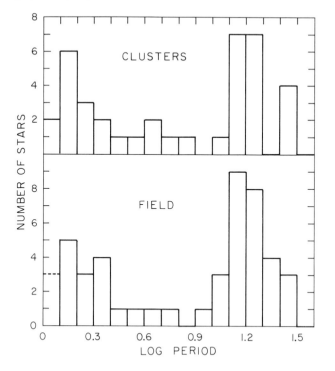

Figure 14.11. Period–luminosity relation for type II Cepheids in globular clusters. Circled dots indicate stars in clusters of type I; dots, stars in clusters of type II.

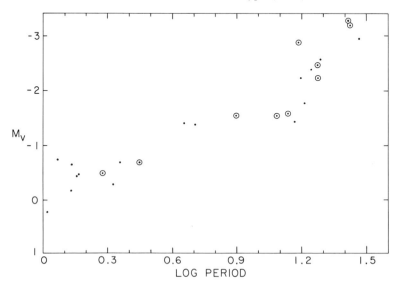

8, stars of a certain mass are expected to execute loops and bends in their tracks of development between the young main sequence and the giant tip; here we find the Cepheids. Theory for the corresponding epoch in the globular clusters is less complete, and the observed arrays themselves more complicated. However, it seems again that stars of a certain (much lower) mass will execute loops in the neighborhood of the red-giant branch and that some of these will cross an instability strip like the one that demarcates the domain of the Cepheids. All these predicted crossings are rapid, in harmony with the observed tendency for the periods of W Virginis stars to fluctuate and vary much more than those of their more massive counterparts.

The W Virginis stars of short period may represent the intrusion of the developing star into the area of instability, as it moves beyond the tip of a blueward-directed stub. Whether it will do this depends on mass, structure, and composition. The group of longer periods may represent stars whose tracks are executing rapid and vigorous loops, crossing and recrossing the instability strip after leaving the second giant branch and losing mass as they do so. Mass, structure, and composition must again be determining factors. This sketch takes account of the two period groups, and brings the short-period W Virginis stars nearer to the RR Lyrae stars, as is also suggested by their periods and luminosities.

If this scenario bears any similarity to reality, we cannot picture the W Virginis stage as either preceding or following the RR Lyrae stage. At the most we can surmise that the short-period W Virginis stars come before the RR Lyrae stars, but the picture is less clear for those of longer period.

Their relation to the blueward end of the horizontal branch is not clear in the color–magnitude diagrams. Still less can these stars be definitely placed in a time sequence. Nor has it been possible to arrange the globular clusters themselves in a time sequence by means of an orderly succession of color–magnitude arrays. The picture presented by observation is confused, and so many parameters are involved that predictions are premature.

The facts are that both RR Lyrae stars and W Virginis stars are found in some clusters. Many clusters rich in RR Lyrae stars contain no known W Virginis stars. A very few, and these of intermediate metal content, have W Virginis stars and no known RR Lyrae stars. None of the thirteen clusters known to include W Virginis stars is of very high metal content, and they are of both type I and type II in regard to the distribution of the periods of their RR Lyrae stars (when these are present). If the surmises of the preceding chapter are correct, their horizontal branches represent both redward and blueward tracks. So we know roughly, but not exactly, where the W Virginis stage falls in the stellar lifetime.

Last Scene of All?

CHAPTER 15

Final stages, like beginnings, are elusive. It is tempting to associate great age with advanced stage of development, but that is an anthropomorphic view. Among living things, senility is not uniquely correlated with great age. The relation for stars is even less clear-cut. What is a senile star? A white dwarf? A pulsar? A black hole? These strike the imagination as having approached the end of the road. But they are not necessarily old in years. Where are senile stars to be found? There are a number of white dwarfs in comparatively young open clusters such as the Hyades, even more in the more advanced open cluster Messier 67. The intermediate-age main-sequence star Sirius, and the not much older Procyon, have white dwarf companions. The motions of the white dwarfs do not associate them uniquely with the halo population that is characteristic of the oldest objects.

It has been necessary, in fact, to distinguish two kinds of age when we considered the life histories of the stars: chronological age, expressed in years, and developmental age, dictated by the stars' inner processes and evinced by condition and behavior. The factor that produces the dichotomy between the two types of age is the mass of the star.

The stars of lowest mass, which have been called the "feather-weights,"* have perhaps less than 7 percent of the sun's mass. They are faint and inconspicuous, although they may be very numerous indeed. In developmental age they are very young: still in the first stage and con-

---

* The nomenclature used in this chapter was suggested by Martin Schwarzschild.

tracting toward the main sequence. But in years they may be as old as the globular clusters.

The "lightweight" stars may have masses up to about four times the sun's. They are very numerous, though not so numerous as the featherweights. Most of the characters that have moved upon our stage are in this class: all the members of globular clusters that have figured in the diagram, and those of the older open clusters. These subsist on a diet of hydrogen and helium; in the time thus far available to them many have reached the red-giant stage. Some become red irregular variables, Mira stars, W Virginis stars, RR Lyrae stars. At and after their fling as red giants they probably lose mass. If the globular clusters are any guide, they finally progress through the blue end of the horizontal branch, and we may envisage an orderly and uneventful progress toward the blue intermediates and finally the white dwarfs.

The "middleweight" stars are more massive: perhaps up to eight or nine times the mass of the sun. These can become blue supergiants and then Cepheid variables before reaching the giant tip. These stars transcend the diet of hydrogen and helium and feed on carbon atoms. Their subsequent careers are anything but uneventful. They may either suffer a complete explosion—a "stellar suicide," as Hubble once described it—or a catastrophic collapse, with violent ejection of their outer parts, leaving behind a neutron star. In either case a supernova would be observed; in the latter case, a pulsar (a radio source with extremely rapid periodic variations) would survive. Stars of such masses would not take long to reach their catastrophe; pulsars may have reached a terminal stage but they need not be old.

In our survey of stellar life we have covered the emergence of of nearly every kind of variable star except the catastrophic variables; we have now sketched the supernovae into our landscape. The novae and dwarf novae have still eluded us, and will not appear until the final chapter. One more type of variable star, the R Coronae Borealis star, may now be mentioned. These stars are exceedingly rare: only a few dozen are known altogether. Their variations consist of sudden and completely erratic drops in brightness—a sort of inverted nova performance. Composition furnishes a clue to their behavior and their origin. Helium and carbon are their dominant constituents; hydrogen is nearly absent. Their sudden drops in brightness can be ascribed to the erratic expulsion of clouds of carbon atoms, which condense to form obscuring envelopes of graphite. Hydrogenless stars are rare indeed. Probably these are light middleweight stars that have blown off their envelopes, leaving their carbon-rich cores exposed to view. It seems unlikely that such a star could last for very long; perhaps it evolves into a planetary nebula. It is understandable that the R Coronae Borealis stars are very rare.

Finally, we consider the fate of the "heavyweight" stars, whose masses are from nine to perhaps a hundred times the mass of the sun. These also reach the point of devouring carbon, and perhaps even heavier nuclei, before their final catastrophe. Here there is probably not an explosion but a collapse, perhaps leading to the formation of a black hole, an aggregation of matter so dense that nothing can escape from it—not even light. Compared to the lighter stars, a single star that becomes a black hole must run its course very fast indeed and may therefore be quite young. But black holes may not originate only from single stars; the well-authenticated Cygnus X-1 is in fact a binary. Today black holes are thought to occur in the cores of globular clusters, probably from the confluence of many stars, and these will therefore be very old chronologically, in harmony with the ages of the clusters in which they are found.

Our "senile stars," then, may be comparatively young in years, though old in experience. But these senior citizens of the stellar community have had a great influence on the life of later generations, for they have enriched the interstellar medium with products of their own stellar digestion. Both the suicidal supernova and the collapsing supernova scatter their heavier elements into space, and the newly-born stars that spring from this material become progressively richer in metals.

The globular clusters and all but the youngest open clusters which have furnished most of our cast of characters exemplify the "lightweight" stars. Here we look for the final stages in the approach to the white dwarf, and we are not disappointed. The Hyades, Coma, and Praesepe all contain white dwarfs; Messier 67 seems to contain over a hundred. The far more populous globular clusters must contain even greater numbers. And in Messier 15 there is at least one planetary nebula.

The planetary nebulae, found in great numbers in the Galaxy, have a distribution and motion that associate them with the disk and the halo. They appear at large distances from the plane and are concentrated toward the galactic center. The nuclei of planetary nebulae are hot white dwarfs, degenerate stars that have exhausted their resources, with a thin fringe of nondegenerate matter. Because the surrounding nebula is expanding and rapidly losing material into the surrounding space, a planetary is an evanescent creature with a lifetime of only a few thousand years. There is little doubt that it represents a very late stage in a career that has perhaps spanned existence as a Mira star, a W Virginis star, or even an R Coronae Borealis star. After the nebulous envelope has been shed, the star will be a white dwarf. Some white dwarfs, therefore, can be relatively young in years, like those in the Hyades. There is a planetary nebula in the open cluster NGC 2818 that is similar to the Hyades in age. Other white dwarfs, toward which the planetary nebula in Messier 15 points the way, are much older. A star can reach senility on many time scales, and by more than one path.

Finale—Pas de Deux

CHAPTER 16

Our scenario has leaned heavily on crowd scenes: large groups of stars, presumably of similar origin, whose interrelationships reveal much about their history. The tracks of development predicted by theory, when converted into isochrones, represent the behavior of these stellar crowds very well. Clusters of stars, rich and poor, conform to a well-defined pattern when their inherited character (with chemical composition the leading factor) is taken into account.

The poorest possible cluster, a cluster with only two members, is a double star. We might expect the two components of a binary system to fall on a single theoretical isochrone, for they must always have been together. That one of them has been captured by the other is wildly improbable. But it turns out that many of the known pairs of stars are extremely unorthodox in their interrelationships. This may seem particularly vexing because they are the stars about which we know the most. They are in fact the *only* stars whose masses are known, and mass is the most vital of all stellar properties. Add to this the fact that (at least in the galactic field) about half the known stars are members of double or multiple systems, and we are compelled to admit that our account of the development of single stars is by no means the whole story. The problem of the mutual influence of stars must be faced. How does the presence of a companion affect stellar development? How far do the tracks of single stars—directed, motivated, and paced by their inner nature—represent those of stars with close companions? Is the behavior of double stars governed by attraction or by physical contact? Is the critical factor gravitation or material interaction? We shall see that the two are closely interwoven.

Most of the double stars we have mentioned have seemed orthodox enough. Besides a narrow, well-defined main sequence, the Pleiades and Hyades show a parallel range of brighter stars that are clearly unresolved binaries (not visible separately) with components that conform to the general pattern. The eclipsing and spectroscopic binaries, like QX Cygni in NGC 7790 and SZ Camelopardalis in NGC 1502 or several of the brighter stars in Messier 7, are not far from the stage they would have reached as single stars. In Praesepe, TX Cancri seems unevolved; as does DS Andromedae in NGC 752; and the short-period eclipsing stars in NGC 188, which must be as old as the cluster, fit the pattern, too.

When we turn to the vast company of known binary stars in the galactic field, we have three sources of information, each evidence of a star's duplicity: visual binaries, pairs of stars observed to be moving in orbits around one another; spectroscopic binaries, whose orbital motion is inferred from periodic changes of radial velocity; and eclipsing binaries, which periodically obscure or conceal one another as they revolve. To be separately observable the components of visual binaries must be far apart and their orbital periods correspondingly long. The known ones range from about two years to centuries, and many are so long that only part of the orbit has been recorded. Spectroscopic binaries, on the other hand, tend to be of shorter period, since (other things being equal), the smaller the orbit, the faster the orbital motion and the easier to detect and measure. Either a visual or a spectroscopic binary could of course perform eclipses if the orbits were suitably oriented, but shorter periods are obviously favored because the two stars are close together. The beauty of eclipses is that they permit an accurate determination of the inclination of the orbit, and therefore of the masses of the components, if both radial velocity curves have been observed.

It is among the visual binaries that we find orthodox pairs, stars that fit the same isochrone. Some of the pairs are virtually identical twins, like the components of γ Virginis (main-sequence F0 stars), others are fainter unevolved main-sequence stars like the two members of ξ Bootis. A more complex but still orthodox group is that of Castor (α Geminorum), which contains three pairs of main-sequence stars, each pair itself a spectroscopic binary and the faintest also an eclipsing system, YY Geminorum (figs. 16.1–16.3).

Main-sequence stars are not invariably teamed up with other main-sequence stars, however. Figure 16.4 illustrates the fact that they can associate with all sorts, from supergiants down to white dwarfs. In many of these pairs the main-sequence star and the evolved star may indeed lie on the same isochrone; we cannot tell unless the masses of the stars are known. But the association of main-sequence stars with white dwarfs suggests a departure from orthodox behavior. Sirius, on the main sequence, Procyon, a subgiant star, and 40 Eridani, a main-sequence G star, all have white-dwarf companions, the latter a member of a binary with a faint main-sequence M component. Most of the white dwarfs in figure 16.4 are teamed up with faint M stars. We must frankly confess that the history of such binary systems is not covered by the simple scheme of isochrones that has served us up to now. A similar problem is posed, of course, by the white dwarfs in the Hyades and other middle-aged clusters. They must represent the remains of initially massive stars that have run their course. It can be argued that the components of these widely separated visual binaries have all pursued their life histories independently, unaffected by their distant companions. Gravitational attraction binds the pairs together and governs their motions, but otherwise the two components exert no mutual influence.

Figure 16.1.   Apparent orbits of γ Virginis (left) and Castor (α Geminorum, right). Their periods of revolution are respectively 177.8 and 380 years.

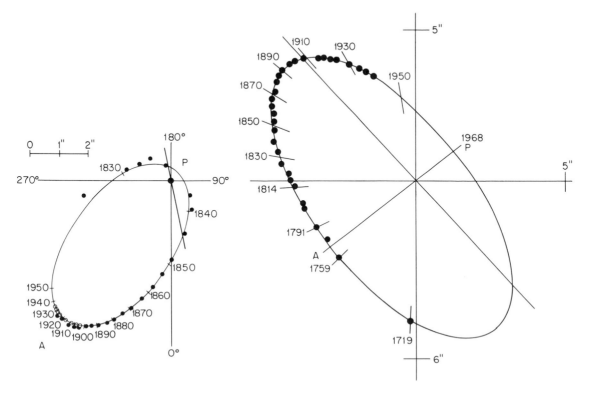

Figure 16.2. Apparent orbit of the visual binary ξ Bootis, easily seen with a small telescope. The stars are of magnitudes 5 and 7. The period is 150 years. (From *Sky and Telescope*, September 1958.)

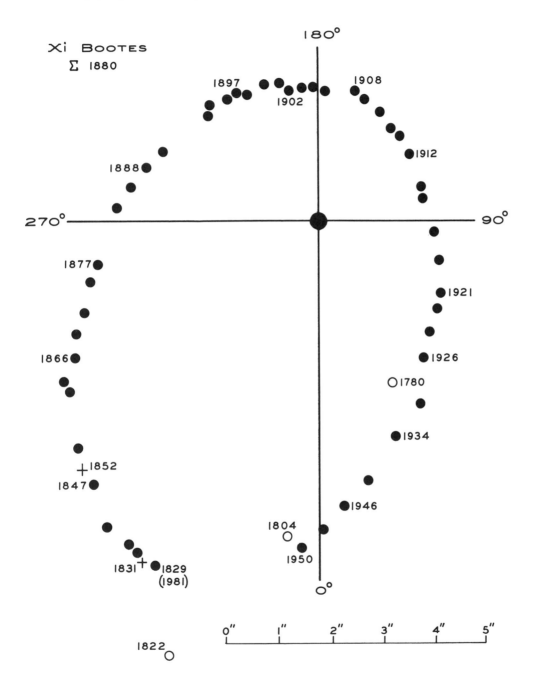

Figure 16.3.   Light curve of the eclipsing binary YY Geminorum, the faint component of the sextuple system of Castor (α Geminorum). The curve shows the variation in 1949, recorded in infrared light. (Based on the work of Gerald Kron.)

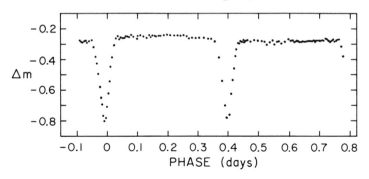

Figure 16.4.   Relation between components of visual binaries. Left: a number of well-known pairs. Right: second components of visual binaries whose first components (not shown in the figure) are main-sequence stars. Crosses are supergiants, circled dots are bright giants, dots are giants, circles are subgiants, small dots are main-sequence stars, and triangles are white dwarfs.

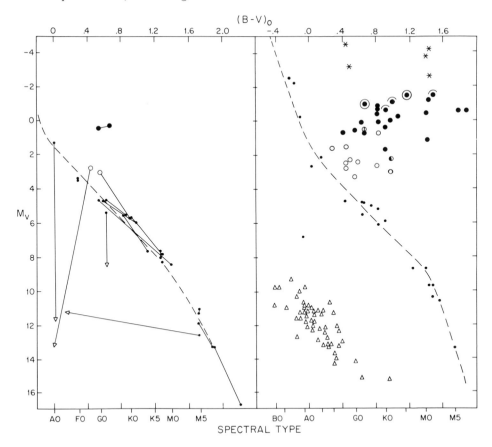

To examine the behavior of closer pairs we turn to the eclipsing binaries, which range from wide separations to pairs almost in contact (table 16.1).

Even among the eclipsing stars of relatively short period we find some orthodox pairs, such as α Coronae Borealis (Alphecca), ZZ Bootis, WZ Ophiuchi, and YY Geminorum (fig. 16.5). The last three systems consist of identical main-sequence twins; the first appears to contain two dissimilar main-sequence stars. All are widely separated and appear to exert little mutual influence. The light variations of the stars just illustrated are of the so-called Algol type, that is, having little or no change of brightness between eclipses. But it would be a mistake to suppose that the components of all Algol stars do not affect one another. The most casual glance reveals evidence of mutual disturbance.

Algol, the Demon star, was probably the first known eclipsing star (fig. 16.6). It is actually multiple; the eclipsing pair is the brighter component —a periodic binary which consists of a hot main-sequence star and a larger, cooler star that is too faint to be observed separately. Because the spectrum of the larger star cannot compete with that of the blue component even in the infrared, the masses of the two stars have not been measured. However, there is little doubt that the fainter one is the more evolved, even though we rely for its dimensions, temperature, and luminosity on the geometry of the eclipse. We should expect the roles to be reversed, and the fainter component to be nearer to the main sequence than its companion. Nor is Algol an isolated exception; it is typical of the large class of Algol stars which have a main-sequence primary and a fainter, more evolved companion.

The bright star U Cephei is another example of an Algol star; here both spectra can be observed, the radial velocities of the components measured, and their masses determined. Figure 16.7 shows the light variations of U Cephei, accurately determined by photoelectric photometry. The deep eclipse marks the passage of the redder star across the face of the bluer; the bluer star is passing across the redder one at the shallow eclipse. Both eclipses are central and we observe the orbit virtually edge-on. The brightness during totality is not quite constant, as would be expected on a crude picture of two uniform spheres passing alternately across one another.

The timing of the eclipses shows that the orbit is circular or very nearly so. In a circular orbit the velocity of each star around the other is always the same, and therefore, the changes in radial velocity, seen in projection as they circulate, will stimulate a sine curve. When the spectroscopic orbit of the blue star was first studied, it came as a surprise that the changes of

Table 16.1. Characteristics of sixteen eclipsing binaries.[a]

Name	Max.	Min.	Spectra	Period (days)	DM	Position h	m	s	°	′
AO Cas	5.96	6.11	O9 III + O9 III	3.52349	+50° 46	0	12	25	+50	52.7
β Per (26 Per)	2.13	3.40	B7.7 V + G8 III	2.86739	+40 673	3	1	40	+40	34.2
ε Aur (7 Aur)	2.94	3.83	F0ep I + ?	9898.5	+43 1166	4	54	47	+43	40.5
ζ Aur (8 Aur)	3.75	3.90	K4 Ib–II + B7	972.176	+40 1142	4	55	29	+40	55.5
VV Ori	5.14	5.51	B1 V + A	1.48538	−1 943	5	28	27	−1	13.6
UW CMa (29 CMa)	4.5	4.8	O8f + O8f	4.3934	−24 5173	7	14	30	−24	22.6
V Pup	4.74	5.25	B1 V + B1 V	1.45449	−48 3349	7	55	22	−48	58.4
α Vir (67 Vir)	0.97	1.04	B1 V + B3 V	4.01416	−10 3672	13	19	55	−10	38.4
α CrB (5 CrB)	2.21	2.32	A0 V	17.3599	+27 2512	15	30	28	+27	3.0
μ¹ Sco	3.0	3.28	B1.5 Vp + B7	1.44027	−37 11033	16	45	6	−37	52.6
μ Sgr (13 Sgr)	3.79	3.92	B8ep Ia	180.45	−21 4908	18	7	47	−21	5.1
β Lyr (10 Lyr)	3.34	4.20	B8p I + B6.5?	12.90814	+33 3223	18	46	23	+33	14.8
υ Sgr (46 Sgr)	4.34	4.44	B8p + F2p	137.939	−16 5283	19	16	0	−16	8.6
V 695 Cyg (41 Cyg)	3.78	3.88	K2 II + B3 V	3784.2	+46 2882	20	10	29	+46	26.3
V 1488 Cyg (42 Cyg)	4.11	4.14	K5 Iab + B4 V	1147.4	+47 3059	20	12	23	+47	24.2
VV Cep	6.65	7.46	M2ep Ia + Be	743.0	+62 2007	21	53	50	+63	9.0

[a] This list contains well-separated pairs (Algol stars) and close, distorted pairs (β Lyrae stars). The former include β Per (Algol), α CrB (Alphecca), μ Sgr, V 695 Cyg, V 1488 Cyg, and VV Cep. The latter include AO Cas, VV Ori, UW CMa, V Pup, μ¹ Sco, and β Lyr. V 695 and V 1488 Cyg are exceptions to the rule that variable stars with Greek letter names do not receive special designations: besides their Flamsteed numbers they are known as o¹ and o² Cygni. Italicized magnitudes are photographic.

Figure 16.5.   Light curves of WZ Ophiuchi, ZZ Bootis, and α Coronae Borealis. (WZ Ophiuchi and α Coronae Borealis from M. G. Fracastoro, comp., *An Atlas of Light Curves of Eclipsing Binaries* [Turin: Osservatorio Astronomico di Torino, 1972]. ZZ Bootis from S. Gaposchkin, *Astronomical Journal* 59 [1954]: 196.)

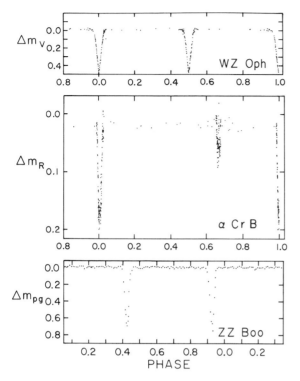

Figure 16.6.   Light curve of Algol (β Persei) in infrared light. (From M. G. Fracastoro, comp., *An Atlas of Light Curves of Eclipsing Binaries* [Turin: Osservatorio Astronomico di Torino, 1972].)

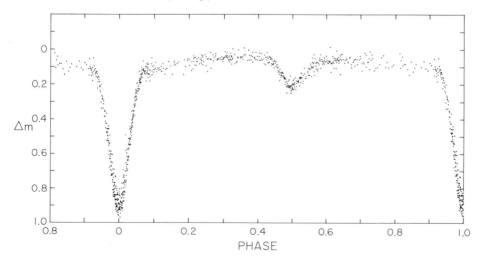

Figure 16.7. Above: light variations of U Cephei in yellow (V), blue (B), and ultraviolet (U) light. Below: sketch of distribution of circumstellar matter in the orbital plane of U Cephei. Densest portion of stream flows directly to the B7 star, within the triangle defined by the phases of the sudden drop in the light curve. The two arrows indicate the streams that could account for the components observed in the lines of ionized calcium and ionized magnesium at the phases shown. Some of the stream returns to the G8 star but probably does not completely encircle the B7 star. Lines of sight to the center of the B7 star at various phases are indicated. (From A. H. Batten, *Publications of the Dominion Astrophysical Observatory*, 14 [1974]: 264, 265.)

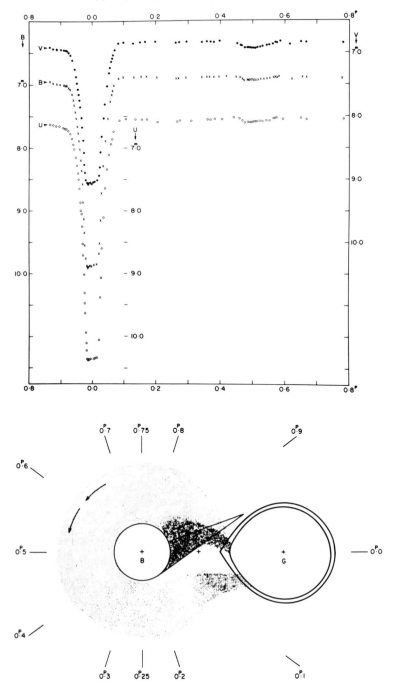

radial velocity did not seem to follow a sine curve but to give evidence of appreciable orbital eccentricity. Controversy raged; U Cephei became a *cause célèbre*; the orbits could not be both circular and eccentric. A critical study of the velocities, and of the structure of the spectral lines on which they were based, finally resolved the mystery. The lines were distorted; they came partly from the blue star, partly from a stream of gases between the stars. The refined picture that emerged was of circular orbits and a stream of gas flowing between the stars and probably also around them.

The components of U Cephei have spectra of classes B7 V and G8 III-IV, so the brighter, hotter star is on the main sequence and the companion is a giant or subgiant. From the radial velocities it follows that the brighter star has 4.2 solar masses, the fainter one 2.8 solar masses. Here there is definite evidence that the less massive of the pair is the more evolved, as has been surmised for Algol, where the masses have not actually been measured.

Figure 16.7 also shows a sketch of the U Cephei system as seen perpendicular to the orbit. Material is flowing from the B star to the G star, with some possibly spilling out into interstellar space as well. This mass loss and mass transfer is steadily changing both components, and it will come as no surprise that each component has not progressed exactly as it would have if the disturbing companion were not present. The orbital period is changing also, probably under the impact of these same conditions; it shows a steady increase as well as small, irregular fluctuations. Even the brightness is inconstant; the minima are not quite flat and are not all exactly alike. And on at least one occasion U Cephei has displayed a small flare in brightness, which was associated with changes in the spectrum and clearly took place in the circumambient gas flow.

Other Algol stars have shown evidence of circumstellar gas flow. Figure 16.8 shows a sketch of the RW Tauri system, where the gas seems to be concentrated in a ring about the blue component, a ring that shows a bright-line spectrum that is eclipsed in the course of the orbital cycle.

Stars like U Cephei and RW Tauri have components that are well separated. The distance between the edges of the two stars in figure 16.7 is comparable to their diameters. The brightness of the system between eclipses is fairly constant. This, in fact, serves as a rough definition of the Algol stars. The other general classes of eclipsing stars, named for the luminous β Lyrae and the low-luminosity W Ursae Majoris, are recognized by a continuous variation between eclipses (fig. 16.9). These stars revolve so close together that their edges are almost touching—they are the so-called "contact binaries." Their mutual influence causes bodily distortion

Figure 16.8.   The eclipsing system RW Tauri. The components are very dissimilar, with specta B8 V and K0 IV; the period is 2.769 days. Above: the complete light curve, from visual observations by Wendell. Below: the bottom of the deep minimum as observed photoelectrically by G. Grant. Five pictures show the relative positions, as seen from Earth, of the B8 (smaller) and K0 stars during the partial and total phases of the eclipse. The spectrum shows that the B8 star is surrounded by a luminous ring that is totally hidden by the K star during the total eclipse. (Based on a study by Alfred H. Joy.)

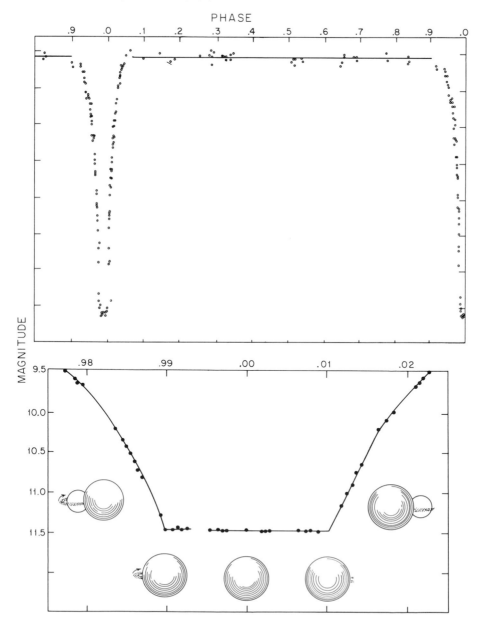

Figure 16.9.   Light curves of W. Ursae Majoris, U Pegasi, and UW Canis Majoris, all contact binaries; the two former are low-luminosity W Ursae Majoris stars; the third is a very luminous β Lyrae star. (W Ursae Majoris and U Pegasi from M. G. Fracastoro, comp., *An Atlas of Light Curves of Eclipsing Binaries* [Turin: Osservatorio Astronomico di Torino, 1972]. UW Canis Majoris from H. Ogata and C. Hukusaku, *Information Bulletin on Variable Stars*, no. 1235 [1977].)

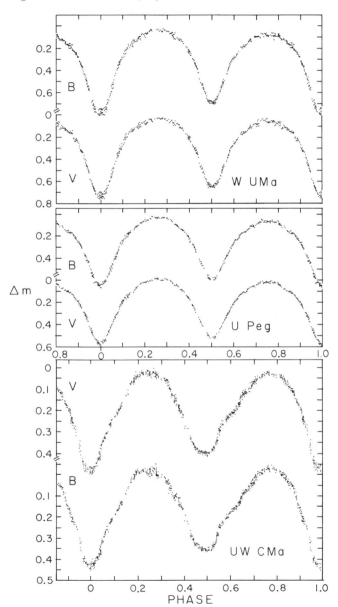

of one or both, so that their joint brightness is affected not only by the eclipses but also by the changing profiles of the stars as they move around one another. Mutual gravitation distorts them into spheroids or even into ellipsoids. Algol stars, on the other hand, are far enough apart for their mutual distortion to play comparatively little part in the variation of brightness outside eclipse.

Studies of their spectra, fortified by theoretical work, have made it clear that these continuously varying contact binaries are not only distorting one another but are also undergoing mass exchange on a greater scale than is observed for the Algol stars. Streams of matter are passing between the stars, and for many of them there is also a well-attested loss of matter into space, so that the total mass of the system is diminishing. Some of them, like SV Centauri, are changing rapidly in period; for this star the period is decreasing very fast, and the components are so close together that they form a kind of dumbbell (fig. 16.10).

The observations of both contact binaries and Algol stars furnish compelling evidence that the life histories of close double stars are profoundly affected by their association. It is the more surprising that a double star such as TX Cancri, which is neatly situated on the main sequence of Praesepe, seems to be unevolved, even though it is a contact binary. Here perhaps material has flowed alternately first from one component, and then from the other and this give and take has a balance that has kept them poised on the main sequence, with no loss of material into interstellar space.

The close binaries among the brightest stars of open clusters, such as DH Cephei and SZ Camelopardalis, are still fairly close to the main sequence. What will be the fate of these luminous, short-period contact binaries? Perhaps they will eventually merge as they move toward the red side of the color–luminosity array, passing through a stage like that of SV Centauri as they go. If they develop along paths appropriate to their rather high masses, will they go the heavyweight route and eventually become supernovae?

Does the mutual influence of close binaries have a bearing on the succession of crucial stages that we have traced in the stellar life history? At the outset we noted that the very youngest stars are found almost invariably in groups or associations, and that as they progress through maturity to old age they tend to be progressively less confined to clusters, though populous clusters may contain large numbers of variable stars. What is the result of an analogous inquiry for the members of binary systems?

Membership in a binary system is certainly not incompatible with the characteristic behavior of a variable star. First among the intrinsic vari-

Figure 16.10.   Three schematic representations of the system of the eclipsing binary SV Centauri, illustrating the high degree of contact. Regions of local temperature excess are shaded and indicated by thicker lines. The period of revolution is decreasing steadily, and more rapidly than for any other known eclipsing star; the components are evidently in active interaction. (From S. M. Rucinski, *Publications of the Astronomical Society of the Pacific*, 88 [1976]: 247.)

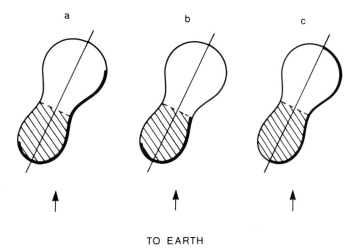

TO EARTH

ables encountered in tracing the stellar history were the β Canis Majoris stars, and a number of these are members of binary systems: β Canis Majoris itself has an orbital period of 49 days, σ Scorpii of 34 days. But the duplicity does not seem to be the cause of the short-period variations that have periods of a few hours. One of the most thoroughly studied of these stars, γ Pegasi (Algenib), is not a binary. And only about a quarter of the naked-eye β Canis Majoris variables are known to be double.

We find a similar situation with the α Canum Venaticorum stars that lie parallel to the main sequence at lower luminosities. The prototype is itself a member of a wide binary of unknown long period; ι Cassiopeiae has an orbital period of 52 years, that of β Coronae Borealis is 10.5 years, and that of ϵ Ursae Majoris is about 4 years. But here again there is no compelling evidence that the characteristic variations are caused by duplicity. Again, only about a quarter of the naked-eye specimens of the class are known to be double.

Next in the sequence of stellar stages that are prone to variability come the Cepheid variables. There are twenty-one known naked-eye Cepheids, and four of them (δ Cephei itself, Polaris, S Sagittae, and FF Aquilae) are members of binary systems. Several fainter Cepheids are also known to be components of double stars, having been recognized as such by perio-

Figure 16.11. Light curve of AB Cassiopeiae. (From M. G. Fracastoro, comp., *An Atlas of Light Curves of Eclipsing Binaries* [Turin: Osservatorio Astronomico di Torino, 1972].)

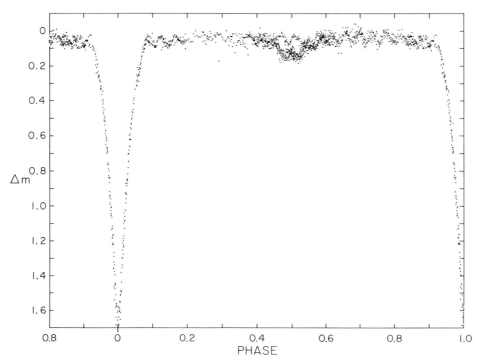

dic changes in their pulsational velocity curves. But by no means all Cepheids, probably less than a quarter of them, are members of binary systems, and therefore their variations cannot be caused by their duplicity.

A similar conclusion can be drawn for the δ Scuti stars, which occupy the lower portion of the Cepheid instability strip. A few are wide binaries, with orbital periods of decades; a few are spectroscopic binaries of shorter period. The most striking is the brighter component of AB Cassiopeiae, an eclipsing binary with period a little over a day (fig. 16.11). But here again only about a quarter of the naked-eye specimens have been found to be components of double stars. There is no reason to conclude that the intrinsic variation is caused by duplicity.

Later in the stellar lifetime comes the Mira star, whose spectacular visual variations endear it to the watcher of the skies. About one in five of those observable with the naked eye are members of binary systems; o Ceti (Mira), R Aquarii, R Hydrae, and X Ophiuchi. Here again, the variations cannot be ascribed to the binary nature of the stars. However, these variables suggest, perhaps for the first time, that one of the members can affect the variation of the other. The orbital periods of Mira, R Hydrae, and X Ophiuchi are long—on the order of centuries or more. That of R Aquarii, though unknown, must be shorter, probably less than a thousand days, so the components will be closer together than for the three other stars. We therefore note with interest that the long-period variation of R Aquarii is at times profoundly affected by the behavior of its companion. The second component of X Ophiuchi appears to be an unremarkable yellow giant star, while those of Mira and R Aquarii are very remarkable blue stars.

Possible binaries among the RR Lyrae stars present an elusive problem. At one time the complex variations of AC Andromedae were ascribed to a system of two RR Lyrae stars. The complex variations of CE Cassiopeiae were in fact shown to arise from the variations of two Cepheids of nearly the same period. Later the two stars were actually observed separately, and the two periods of CE Cassiopeiae a and b were confirmed. But for AC Andromedae the more probable interpretation involves an extreme case of the complex variations of a single star, a not uncommon thing for RR Lyrae variables.

The brief survey of intrinsic variables as components of double stars suggests that most of them are governed by their inner nature rather than by the proximity of a companion. The behavior of the pair is dictated by attraction rather than by physical contact, and on the whole they are well separated—AB Cassiopeiae may be an exception. But for the very large stars, the Mira variables, there is a suggestion of physical interaction as well. Other large, cool stars—the red giants and supergiants, components of such eclipsing systems as VV Cephei, AR Pavonis, and CI Cygni— while varying in their own characteristic manner, also furnish evidence of mutual influence. But there are very many such stars that are not members of binaries.

In fact, we conclude that although most kinds of intrinsic variables can be found among the components of double stars, duplicity is not the cause of their characteristic behavior. A large proportion of all stars are members of double or multiple systems and in this respect intrinsic variables are no exception.

The aficionado of variable stars will have looked in vain in our account for one of the greatest of all stellar exploits, the cataclysmic event shown variously by the "classical" nova and the dwarf nova or U Geminorum star. Novae are not as other stars. Their outbursts do not stem from their inner nature but are associated with the presence of a close companion. Novae are the only variable stars that are always binary systems.

Figure 16.12. Light curves of fifteen typical novae, all drawn to the same scale. The diagram illustrates the variety of development. All novae increase rapidly in brightness, but some decline rapidly, others slowly; some develop smoothly, others show large oscillations of brightness. Ordinates are marked at intervals of one magnitude, abscissae at intervals of ten days.

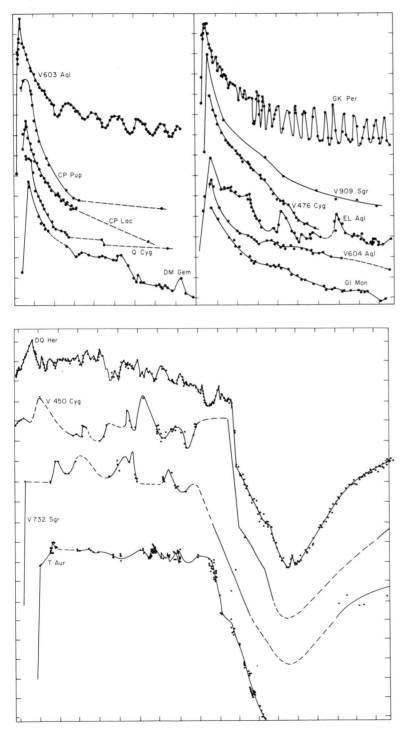

Classical novae increase abruptly in brightness, typically by about thirteen magnitudes but occasionally, as for V 1500 Cygni 1975, by more than eighteen (fig. 16.12). There may be a premonitory rise before the main outburst. Their spectra reveal a rapidly expanding envelope, rising with velocities of several thousand kilometers a second for the fastest to perhaps a hundred kilometers a second for the most leisurely. At first the spectrum looks like that of a high-luminosity supergiant, and soon develops in a complex succession of absorption lines that give way sooner or later to bright-line spectra, first with permitted, then with forbidden, lines. There has evidently been some sort of violent ejection of material, which always gives evidence of axial symmetry. Typically (though perhaps not always) the nova returns to its pre-outburst brightness. I resist the temptation to describe the exquisite details any further.

The early study of novae was understandably directed toward an analysis of the explosion, of the physical changes in the expanding envelope, and of the geometry of the ejecta. Gradually it became clear that while the outbursts showed great variety (rapid and slow, smooth and fluctuating in brightness), after they reverted to the pre-outburst condition novae had much in common. They were faint blue stars, sometimes with bright lines in their spectra and sometimes with evidence of a component of late spectral type.

The dwarf novae or U Geminorum and Z Camelopardalis stars seemed to differ more from the novae than the novae differed among themselves. Their outbursts were of smaller range and occurred repeatedly in cycles of weeks or months. The spectra never developed the complex succession of absorptions that are found for all novae, nor did they display bright lines at maximum, either permitted or forbidden. But they had this in common with novae: at minimum they showed the spectra of faint blue stars, sometimes with evidence of a late component. The blue stars seemed of lower temperature than the post-novae, but otherwise they were similar.

The first evidence of binary nature for cataclysmic variables emerged when the U Geminorum star SS Cygni was shown to be a spectroscopic binary with a period of about a quarter of a day, a system consisting of a very blue star and a main-sequence G5 star not unlike the sun (fig. 16.13). Compelling evidence of binary nature for a classical nova was furnished by the discovery that Nova DQ Herculis (1934) is an eclipsing binary with a period about two tenths of a day, and this discovery was followed by the demonstration that it is a spectroscopic binary. We now know that three classical novae (RR Pictoris, DQ Herculis, and T Aurigae) are eclipsing systems and four (V 603 Aquilae, HR Delphini, DQ Herculis, and GK

Figure 16.13. Variations of two dwarf novae, SS Cygni (above) and Z Camelo-pardalis (below). The former has continuous cyclic outbursts, the latter is less re-petitive and undergoes intervals of quiescence. (From observations made by the American Association of Variable Star Observers and by Luigi Jacchia.)

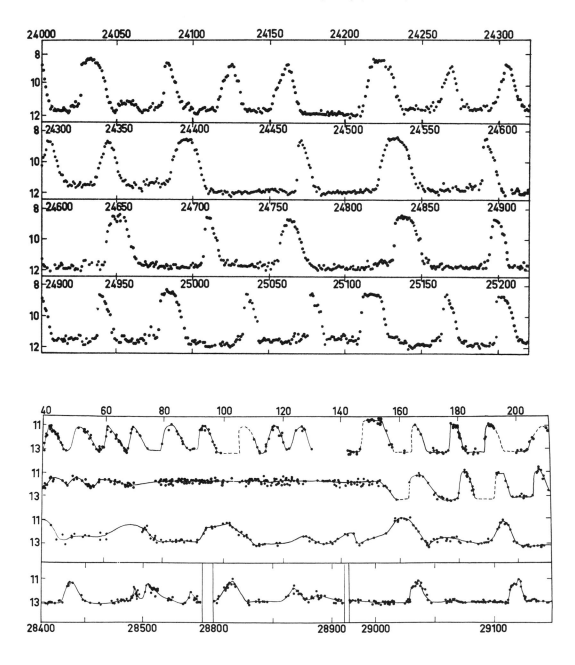

Persei) are spectroscopic binaries. Six dwarf novae are eclipsing stars, and nine are spectroscopic binaries. The accumulating evidence leads to the conclusion that all novae and dwarf novae are double stars, and that they are essentially similar systems, consisting of a faint blue star and a late main-sequence component. Figure 16.14 shows eclipses of DQ Herculis (a nova), U Geminorum (a dwarf nova), and UX Ursae Majoris (a potential nova). Figure 16.15 shows a schematic model that represents the physical makeup of a nova. Since the second outburst of T Pyxidis in 1902 (it was first observed as a nova in 1890), we have known that a nova can erupt more than once; this nova, in fact, has had subsequent outbursts in 1920, 1944, and 1970, all essentially similar in character. Nova T Coronae Borealis, first observed in 1866, erupted again in 1946, and several other recurrent novae are recorded. They furnish a link between the classical novae and the dwarf novae: their ranges of brightness are intermediate, and their cycles (decades as compared with tens of days) are longer than those of dwarf novae. Whether all novae are recurrent is an open question: cycles of several centuries can have passed unnoticed, and all novae (especially those of small observed range) are worth watching in the hope of reappearance.

There seems to be a real difference between the classical novae that have been shown to be binaries of short period and the recurrent novae. Whereas the former consist of blue stars and faint main-sequence companions, most of the latter are binaries that contain a blue star and a giant companion. Nova T Coronae Borealis, a spectroscopic binary with an orbital period of 228 days, consist of a faint blue star and a giant M companion. Novae may all be binaries, but not all are of the same kind. What have they in common? The outburst of T Coronae Borealis, though perhaps unusually violent, was not unlike those of some classical novae.

A by-product of the study of novae and dwarf novae at minimum has been the discovery of a number of binary systems that are very similar to them but that have never been associated with a major outburst. These range from relatively quiescent eclipsing systems like UX Ursae Majoris and RW Trianguli to obviously disturbed stars that vary spasmodically, like VV Puppis, TT Arietis, VZ Sculptoris, and EM Cygni, all of which are both eclipsing and spectroscopic binaries. We need not hesitate to regard these stars as close relatives of the novae; perhaps they have suffered a major outburst in the past, perhaps are preparing for one in the future, perhaps both. Nova outbursts in our galaxy are not uncommon; if they have occurred in the past at the current rate, there must be thousands, perhaps tens of thousands, of ex-novae within reach of our observations.

Figure 16.14.   Light curves of eclipses of DQ Herculis, UX Ursae Majoris, and U Geminorum. (DQ Herculis from Merle Walker, *Astrophysical Journal*, 127 [1958]: 319. UX UMa from H. Johnson, B. Perkins, and W. A. Hiltner, *Astrophysical Journal*, 1 [1954]: 91. U Geminorum from G. Mumford, *Astrophysical Journal*, 139 [1964]: 476.)

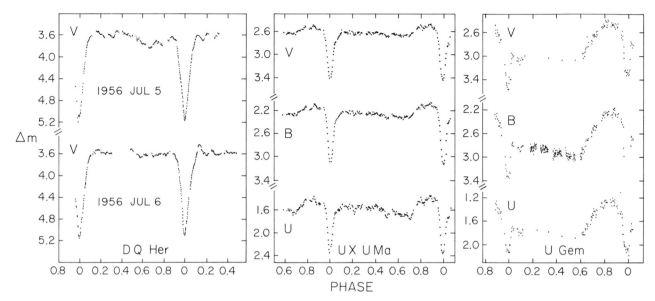

Figure 16.15.   The basic model for cataclysmic variables. The geometry pictured here corresponds to the dwarf nova Z Cameloparadalis. The direction of orbital motion is counterclockwise. (From E. L. Robinson, *Annual Review of Astronomy and Astrophysics*, 14 [1976]: 121.)

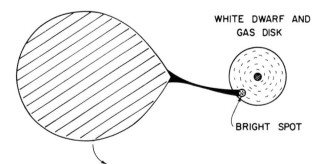

Or perhaps some of these nova-like systems may be fated to simmer continuously without ever achieving a major eruption.

The accumulating evidence that cataclysmic variables are binary systems whose essential ingredient is a faint blue star brings us a little nearer to the crucial question: what causes the nova outburst? The discovery of recurrent novae disposes of the old idea that the event is a final catastrophe. The very commonness of the nova phenomenon rules out the possibility of an isolated accident such as a stellar collision. And the conviction that novae are binaries leads inescapably to the conclusion that here, at last, we see the full effects of physical contact, of material interaction. That such effects can occur and can profoundly affect stellar development is already evident for the Algol stars, still more so for the W Ursae Majoris stars, the contact binaries.

Cataclysmic variables are contact binaries of a special kind. The essential ingredients are a white dwarf and another star (a main-sequence star for the binaries of short period, a giant star for long-period systems like T Coronae Borealis) so placed that material is gravitationally constrained to flow from it toward the white dwarf. This material accumulates in a disk that circulates around the white dwarf in the plane of the binary orbit, and the point of impact of the flow upon the disk is the site of a shock front that appears as a bright spot. The white dwarf is hidden by the circumambient flow of material; its presence is inferred from the fact that rapid oscillations (with periods of tens of seconds) are seen in the light of the system, transmitted by the gases in the disk (fig. 16.16). The periods of these oscillations are appropriate to a body of the density of a white dwarf and are regarded as evidence of its presence, though it rarely if ever can be observed directly.

This extremely circumstantial account of the structure of the cataclysmic binary is no flight of fancy. It is based on delicate studies of the variations of brightness and spectrum too technical to be described here. The light curves of the periodic cataclysmic variables are the outcome of the interplay of disk and bright spot, one or the other of which dominates at different phases in accordance with the geometry of the system. If the flow of material is spasmodic, the intensity of the bright spot flickers, and this effect is responsible for the very rapid and erratic variations that take place on a time scale of seconds and minutes. Each system presents its own continuous evidence of physical interaction. Figure 16.17 shows the spasmodic variations of GK Persei after its decline from the outburst of 1901.

This sketch of the cataclysmic variable at minimum, built up from the details of observation, unites the classical novae, the dwarf novae, and

Figure 16.16.   Rapid intrinsic variations of DQ Herculis, evidence of the oscillations of the white-dwarf component. (From Merle Walker, *Astrophysical Journal*, 123 [1956]: 81.)

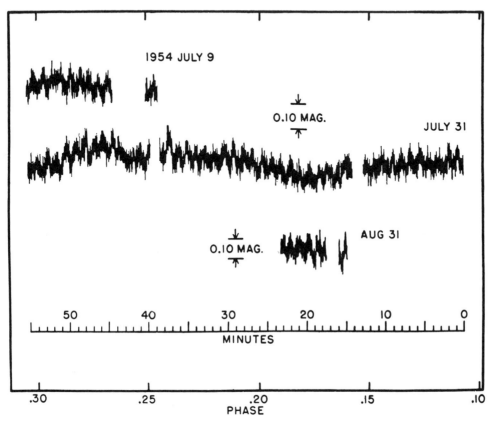

nova-like variables or potential novae. For a full understanding of the problem, the model must be related to the great variety of observed outbursts, from classical novae through recurrent novae and dwarf novae to the systems that have never had a major eruption. The orbital periods of the binaries seem quite unrelated to the violence of the outburst or to the length of the outburst cycle. There is a growing belief that the source of the explosion for a classical nova is a "runaway" outburst of thermonuclear energy provoked in the envelope of the white dwarf, probably involving accreted material in the disk. Such an explosive hydrogen-burning event requires that the white dwarf be anomalously rich (by solar standards) in carbon, nitrogen, and oxygen. Chemical analysis of the ejecta of novae is extremely tricky, but the available results seem to point to just such an enrichment. The nova outburst may well be evidence of

Figure 16.17.   Variations of GK Persei (the bright nova of 1901) after its return to minimum light. Julian days cover the interval 1921 through 1932. The variations are large and erratic. (Based on the visual observations of W. H. Steavenson and B. M. Peek.)

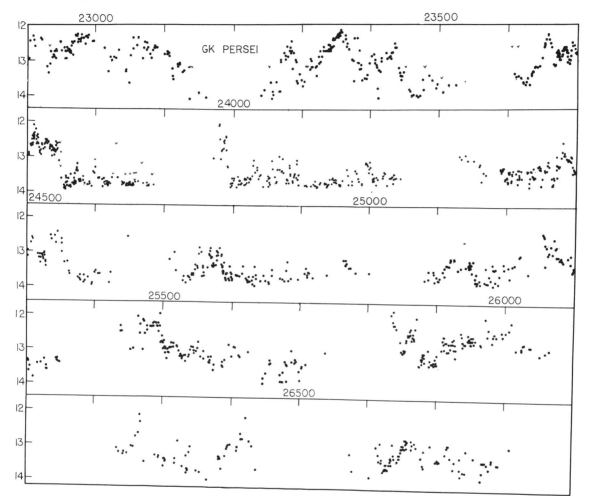

physical contact between stars provoked by gravitational attraction and carried to the extreme limit.

This process, however, is too drastic for the dwarf novae, which appear to swell up at intervals rather than to erupt. As for the classical novae, the white dwarf seems to be the seat of the outburst. In some cases, perhaps in all, the brightening may be occasioned by changes in the disk, possibly abetted by the white dwarf itself.

The picture for the recurrent novae may be analogous, but a red giant must be substituted for the main-sequence star. In some ways this makes things easier, for many, perhaps most, red giants are spilling material into space, so that interaction between the two stars is very probable. But a detailed model has not been formulated as yet.

The cataclysmic variables during outburst impress us with their diversity: classical novae of both rapid and slow development, recurrent novae, dwarf novae, quiescent nova-like stars. At minimum luminosity they impress us with their similarity—all are camouflaged white dwarfs with material streaming to them from low-temperature companions. But each system has its limitations, its restricted repertoire. Recurrent novae repeat their outbursts with some fidelity. The U Geminorum and Z Camelopardalis stars repeat their mannerisms in irregular but recognizable cycles. Each system is true to itself. Beneath the now familiar system of white dwarf, red companion, disk, and spot there lies an individuality to which as yet we have no clue.

The recurrent novae provide a link with a large and diverse group of stars that have come to be known as the "symbiotic variables" (a name that, if it were not already preempted, would have been appropriate to the cataclysmic variables just described). Three well-known examples are AG Pegasi, BF Cygni, and CI Cygni (fig. 16.18). The first consists of a Wolf-Rayet star and a giant M star, the two last of bright-lined B and giant M stars. All are spectroscopic binaries with periods between 750 and 850 days. The blue components have masses of the same order as those of the blue components of the cataclysmic binaries, and all have a tendency to variability. Besides being a spectroscopic binary, CI Cygni is an eclipsing star and has undergone at least one nova-like outburst. If T Coronae Bo-

Figure 16.18.   Average light curve of the eruptive binary CI Cygni, which is an eclipsing variable with an orbital period of 855 days. (Based on a study by T. S. Belyakina.)

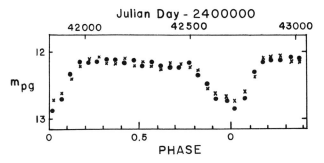

realis is the analogue of a classical nova, perhaps these stars are the analogues of U Geminorum stars, with giant companions. That they are in some sense cataclysmic variables cannot be doubted. Two other stars in the same class not yet known to be spectroscopic binaries are Z Andromedae and AX Persei; both combine a blue star with a giant M star, and both have undergone recurrent nova-like outbursts (fig. 16.19). These stars are surely examples of physical interaction with cataclysmic results.

We turn finally to the Mira stars with peculiar blue companions, Mira Ceti and R Aquarii (fig. 16.20). The latter must be classed as a true cataclysmic variable. In 1921 the blue component underwent an outburst; its continuum dominated the spectrum for several months, and at the same time the characteristic long-period variation of the red star was almost suppressed, recovering only as the blue component faded—a striking example of interaction between the components. There has evidently been another outburst in the far past, for the star is surrounded by an expanding nebula and the forbidden lines of O III are still visible in its spectrum.

The case of o Ceti (Mira) is different (fig. 16.21). The blue companion VZ Ceti is far more distant from Mira than the companion of R Aquarii is from that star, but its peculiar spectrum and its rapid intrinsic variations recall those of the blue components of novae. It is hard to resist the impression that material ejected from Mira in the course of its pulsations is reaching and affecting its companion.

It has been possible to isolate a number of crucial epochs in the career of a star, and in some sort to relate them to the timetable of stellar development; most stars have found their places in the pattern. But some double stars, particularly the close pairs, have conspired to defeat our efforts. The theory of stellar development is all very well, it has been said, until we try to apply it to double stars. Then we must proceed with caution.

To the question of where novae are to be placed in the stellar time scale we must return an equivocal answer. When we appeal to the touchstone of distribution, we find that the novae are dispersed through the Galaxy in a somewhat flattened disk population. They are noticeably concentrated toward the galactic center, but their distribution does not identify them with a halo population. Few are at great distances from the galactic plane. They do not seem to be uniquely associated with any one type of stellar population.

It is true that three novae have been suggested as possible members of globular clusters, which would place them in the halo. In 1860 T Scorpii appeared in Messier 80 (NGC 6093); V 1148 Sagittarii appeared in 1943 near NGC 6553 (but not near enough to make it an undoubted member);

Figure 16.19. Variations of the eruptive binary Z Andromedae made photographically in blue light (dots) and in yellow light (circles). These data show that the bright episodes involve the blue component. (From C. Payne-Gaposchkin and C. Boyd, *Astrophysical Journal*, 104 [1946]: 363.)

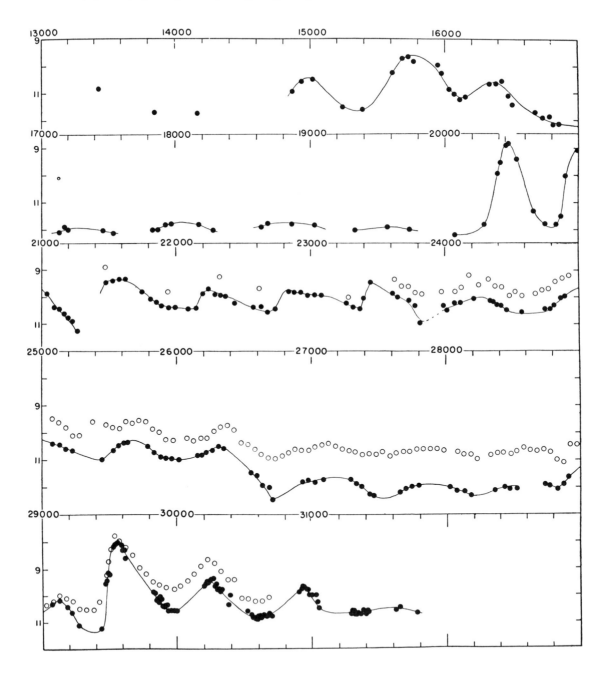

Figure 16.20   Observations of the Mira variable R Aquarii in blue light (dots) and yellow light (circles). The supperssion of the long-period variations during the interval when the blue component was relatively bright is clearly seen between Julian Day 26000 and 27000 (1930 and 1933). (From C. Payne-Gaposchkin and C. Boyd, *Astrophysical Journal*, 104 [1946]: 358.)

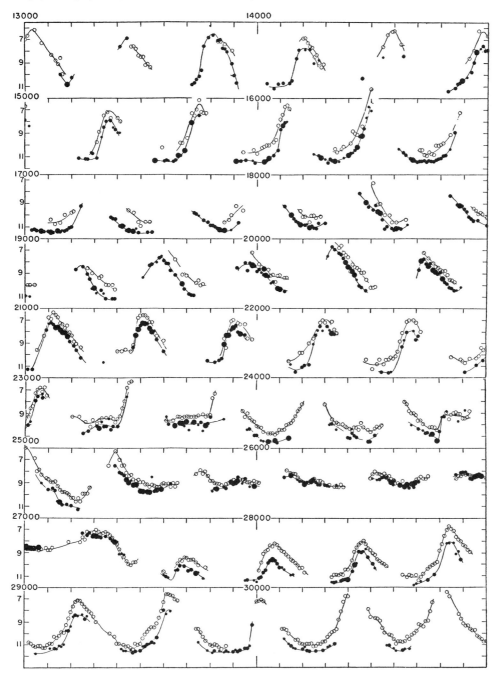

Figure 16.21   The rapid irregular variations of the system of Mira Ceti during an interval of a little more than half an hour; they must be ascribed to the faint blue companion, VZ Ceti. (From Brian Warner, *Monthly Notices the Royal Astronomical Society*, 159 [1972]: 95.)

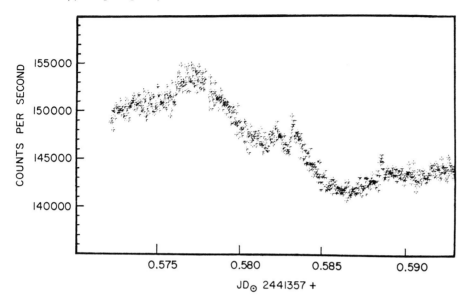

and a nova occurred in Messier 14 (NGC 6402) in 1938. Three dwarf novae have also been associated with globular clusters: one in Messier 5 (NGC 5904), one in NGC 6712, and one in Messier 30 (NGC 7099). The incidence of these six stars, unlike that of the Mira stars or the RR Lyrae stars, does not seem to be related to the metallicity of the clusters or to the character of their color–magnitude arrays. On the other hand, a dwarf nova has been suggested as a possible member of NGC 2482, an open cluster not unlike the Hyades in makeup and age. If BX Puppis is indeed a member of this relatively young system, it suggests that a nova is not necessarily old.

Very few double stars have been found in the vicinity of globular clusters, less than half a dozen eclipsing stars in all, and some of these do not share the motion of the clusters. Considering the intensive search that has been made for variable stars in these systems, we may infer that double stars are actually rarer there than in open clusters or the galactic field, where W Ursae Majoris stars are very numerous indeed. Many years ago, Walter Baade made a search expressly designed to find W Ursae Majoris stars among the faintest accessible stars in Messier 3. His verdict (which I do not believe he ever published) was that there are none. Another point to be remembered is that globular clusters are much poorer in metals than are most stars in the galactic field; the current theories of the nova outburst require that novae be rich in carbon, nitrogen, and oxygen, which might imply that they would not be poor in metals.

In a search for the ancestors of novae, a case has been made for the idea that W Ursae Majoris stars are their progenitors. These eclipsing stars are very numerous; indeed, when allowance is made for their low luminosity and the small volume of space in which they have been surveyed, they appear to be more numerous, by a factor of ten to thirty, than all other types of eclipsing stars combined. We have seen that, as contact binaries, they are subject to continuous mass transfer, and it is difficult to assign a time to their development or to predict with any certainty the mutual transformations they are destined to undergo. Perhaps the only thing we can say is that the nova phenomenon is something that can overtake any binary that consists of a white dwarf and a main-sequence star, provided the components are close enough together to permit active physical contact.

We have discussed the interaction of pairs of stars at some length. Wider vistas are opened up when the interaction of whole stellar communities is envisaged. From a consideration of the dynamic evolution of globular clusters, it has been concluded that within 10^{10} years their central cores must inevitably collapse, and massive supernovae will probably result. Are we witnessing premonitory symptoms of such an occurrence in the recent observations of outbursts of X rays at the centers of some globular clusters in the nucleus of our galaxy? Such an event, a massive supernova, would be more spectacular than anything we have yet witnessed on the restricted stage of our own galaxy.

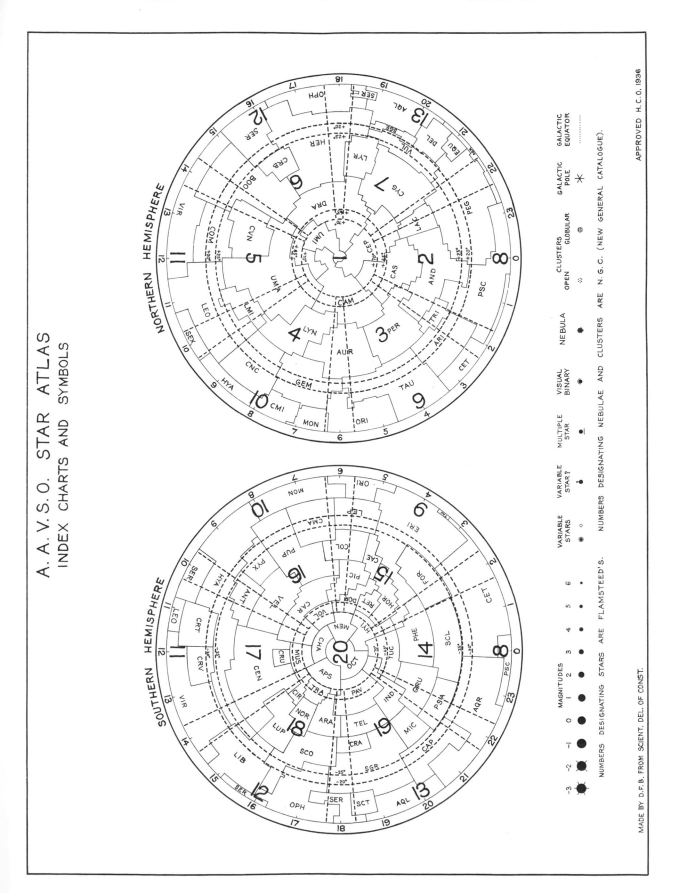

A.A.V.S.O. STAR ATLAS

INDEX CHARTS AND SYMBOLS

NORTHERN HEMISPHERE

SOUTHERN HEMISPHERE

MAGNITUDES

-3 -2 -1 0 1 2 3 4 5 6

NUMBERS DESIGNATING STARS ARE FLAMSTEED'S.

MADE BY D.F.B. FROM SCIENT. DEL. OF CONST.

VARIABLE STARS

VARIABLE STAR?

MULTIPLE STAR

VISUAL BINARY

NEBULA

CLUSTERS

OPEN   GLOBULAR

GALACTIC POLE

GALACTIC EQUATOR

NUMBERS DESIGNATING NEBULAE AND CLUSTERS ARE N.G.C. (NEW GENERAL CATALOGUE).

APPROVED H.C.O. 1936

A.A.V.S.O. STAR ATLAS

CHART I

EPOCH 1900

CHARACTERISTICS OF BRIGHT STARS

ALPHA ANDROMEDAE
MAGNITUDE 2.15
SP. BINARY, PERIOD 96.7 D.
PARALLAX 0".024, 136 L.Y.
INTR. LUMIN. 209 × SUN
P. MOTION +0".143, −0".163

ALPHA CASSIOPEIAE
IRR. VAR. MAGN. 2.1 − 2.6
PARALLAX 0".011, 296 L.Y.
MAX. INTR. LUM. 1040 × SUN
P. MOTION +0".053, −0.033

ALPHA URSAE MIN.
CEPH. VAR. MAGN. 2.1 − 2.2
PERIOD 3.968577 DAYS
VIS. BIN. WITH 8ᵐ.6, PER.—
PARALLAX 0".003, 1086 L.Y.
MAX. INTR. LUM. 13.978 × SUN
P. MOTION +0".043, +0".001

ALPHA ERIDANI
MAGNITUDE 0.60
PARALLAX 0".049, 66 L.Y.
INTR. LUMIN 209 × SUN
P. MOTION +0".088, −0".031

GAMMA ANDROMEDAE
MAGN. 2.20 (2.28 : 5.39 : 6.59)
BC VIS. BIN. PER. 55 Y.
A − BC DIST. 10", P0S. 64°
PARALLAX 0".005, 652 L.Y.
INTR. LUMIN. 4589 × SUN
LUMINOSITY RATIO 1738 : 1
P.M. A +0".049, −0".054
P.M. BC +0".048, −0".057

OMICRON CETI
MAGN. (2.0 − 10.1) : 10.0
L.P. VAR. PERIOD 332 D. ±
A SP. BIN. PERIOD 331 D.
AB VIS. BIN. PER. —
PARALLAX 0".013, 251 L.Y.
MAX. INTR. LUM. 816 × SUN
MAX. DIAM. A 461 × SUN
MAX. VOL. A 97 923 000 × SUN
P. MOTION +0".002, −0".239

BETA PERSEI
ECL. BINARY, MAGN. 2.2 − 3.5
TYPICAL ALGOL VAR.
PERIOD 2.8673I016 DAYS
SP. TRIPLET, SEC. PER. 1.87 Y.
PARALLAX 0".031, 105 L.Y.
MAX. INTR. LUM. 119 × SUN
DIAMETER A 1.22 × SUN
VOLUME A 1.8 × SUN
P. MOTION +0".008 , −0".007

ALPHA PERSEI
MAGNITUDE 1.90
PARALLAX 0".009, 362 L.Y.
INTR. LUMIN. 1867 × SUN
P. MOTION +0".028, −0".030

ALPHA TAURI
MAGNITUDE 1.06
COMP. 13ᵐ, DIST. 31", P0S. 109°
PARALLAX 0".046, 71 L.Y.
INTR. LUMIN. 155 × SUN
DIAMETER 47 × SUN
VOLUME 100 700 × SUN
P. MOT. AB +0".071, − 0".192

EPSILON AURIGAE
ECL. VAR.? MAGN. 3.1 − 3.8
PERIOD 27.059 YEARS
PARALLAX 0".002, 1629 L.Y.
MAX. INTR. LUM. 12.521 × SUN
P. MOTION +0".006, −0".014

ALPHA AURIGAE
MAGN. 0.21 (0.8 : 1.1)
SP. BIN. PER. 104.022 D.
RESLV'D. BY INTERFEROMETER
PARALLAX 0". 068, 48 L.Y.
INTR. LUMIN. 155 × SUN
DIAMETER A 12 × SUN
VOLUME A 1728 × SUN
P. MOTION +0".088, − 0".430

BETA ORIONIS
MAGN. 0.34 (0.34 : 8.7 : 8.7)
A SP. BIN. PER. 21.9 DAYS
BC VIS. BIN. PER. —
A− BC DIST. 9", P0S.201°
PARALLAX 0".006, 543 L.Y.
INTR. LUM. 17 676 × SUN
P. MOT. A +0".005, −0".002

GAMMA ORIONIS
MAGNITUDE 1.70
PARALLAX 0".017, 192 L.Y.
INTR. LUMIN. 629 × SUN
P. MOTION −0".004, −0".019

BETA TAURI
MAGNITUDE 1.68
PARALLAX 0".022, 148 L.Y.
INTR. LUMIN. 383 × SUN
P. MOTION +0".034, −0".177

EPSILON ORIONIS
MAGNITUDE 1.75
PARLLX. < 0".001?, >3258 L.Y.?
INTR. LUM. > 173 660 × SUN?
P. MOTION +0".003, −0".002

(CONTINUED ON CHART 2)

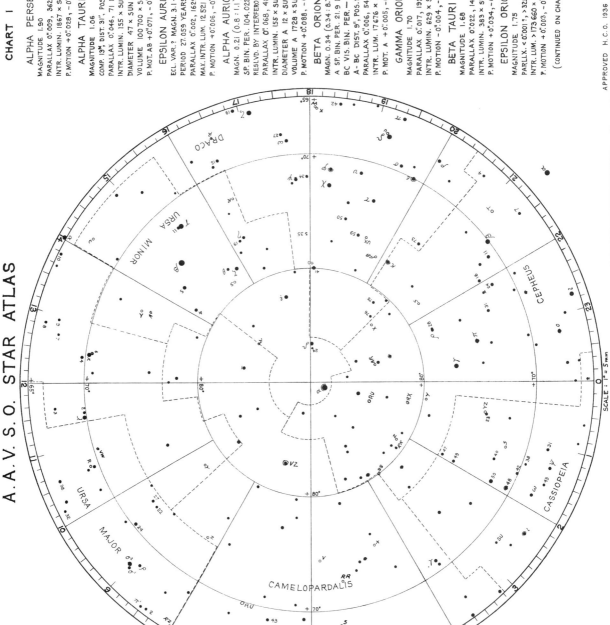

SCALE 1" = 5 mm

MADE BY D.F.B. FROM R.H.P.

APPROVED H.C.O. 1936

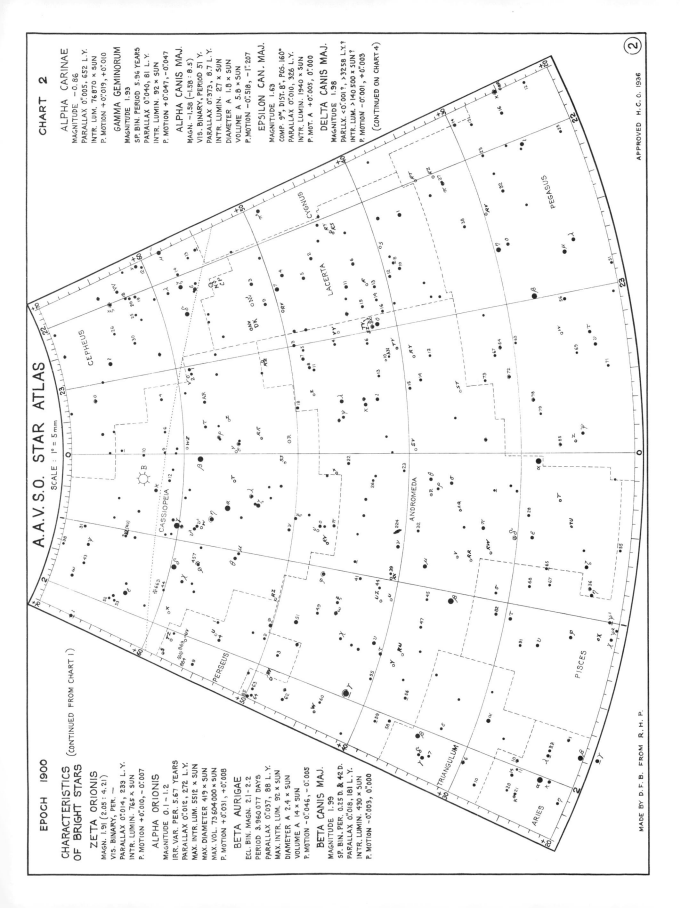

A.A.V.S.O. STAR ATLAS

SCALE : 1° = 5 mm

EPOCH 1900

CHARACTERISTICS OF BRIGHT STARS (CONTINUED FROM CHART 1)

ZETA ORIONIS
MAGN. 1.91 (2.05 : 4.21)
VIS. BINARY, PER. —
PARALLAX 0."014, 233 L.Y.
INTR. LUMIN. 765 × SUN
P. MOTION +0."010, −0."007

ALPHA ORIONIS
MAGNITUDE 0.1 – 1.2
IRR. VAR. PER. 5.67 YEARS
PARALLAX 0."012, 272 L.Y.
MAX. INTR. LUM. 5512 × SUN
MAX. DIAMETER 419 × SUN
MAX. VOL. 73 604 000 × SUN
P. MOTION +0."031, +0."008

BETA AURIGAE
ECL. BIN. MAGN. 2.1 – 2.2
PERIOD 3.960 077 DAYS
PARALLAX 0."037, 88 L.Y.
MAX. INTR. LUM. 92 × SUN
DIAMETER A 2.4 × SUN
VOLUME A 14 × SUN
P. MOTION −0."046, −0."005

BETA CANIS MAJ.
MAGNITUDE 1.99
SP. BIN. PER. 0.25 D. & 42 D.
PARALLAX 0."018, 181 L.Y.
INTR. LUMIN. 430 × SUN
P. MOTION −0."003, 0."000

ALPHA CARINAE
MAGNITUDE −0.86
PARALLAX 0."005, 652 L.Y.
INTR. LUM. 76 870 × SUN
P. MOTION +0."019, +0."010

GAMMA GEMINORUM
MAGNITUDE 1.93
SP. BIN. PERIOD 5.96 YEARS
PARALLAX 0."040, 81 L.Y.
INTR. LUMIN. 92 × SUN
P. MOTION +0."047, −0."047

ALPHA CANIS MAJ.
MAGN. −1.58 (−1.58 : 8.5)
VIS. BINARY, PERIOD 51 Y.
PARALLAX 0."373, 8.7 L.Y.
INTR. LUMIN. 27 × SUN
DIAMETER A 1.8 × SUN
VOLUME A 5.8 × SUN
P. MOTION −0."518, −1."207

EPSILON CAN. MAJ.
MAGNITUDE 1.63
COMP. 9ᵐ, DIST. 8", POS. 160°
PARALLAX 0."010, 326 L.Y.
INTR. LUMIN. 1940 × SUN
P. MOT. A +0."005, 0."000

DELTA CANIS MAJ.
MAGNITUDE 1.98
PARLLX. <0."001?, >3258 L.Y.?
INTR. LUM. >140 500 × SUN ?
P. MOTION −0."001, +0."003
(CONTINUED ON CHART 4)

CHART 2

APPROVED H.C.O. 1936

MADE BY D.F.B. FROM R.H.P.

CEPHEUS
CASSIOPEIA
PERSEUS
CYGNUS
LACERTA
ANDROMEDA
PEGASUS
PISCES
TRIANGULUM
ARIES

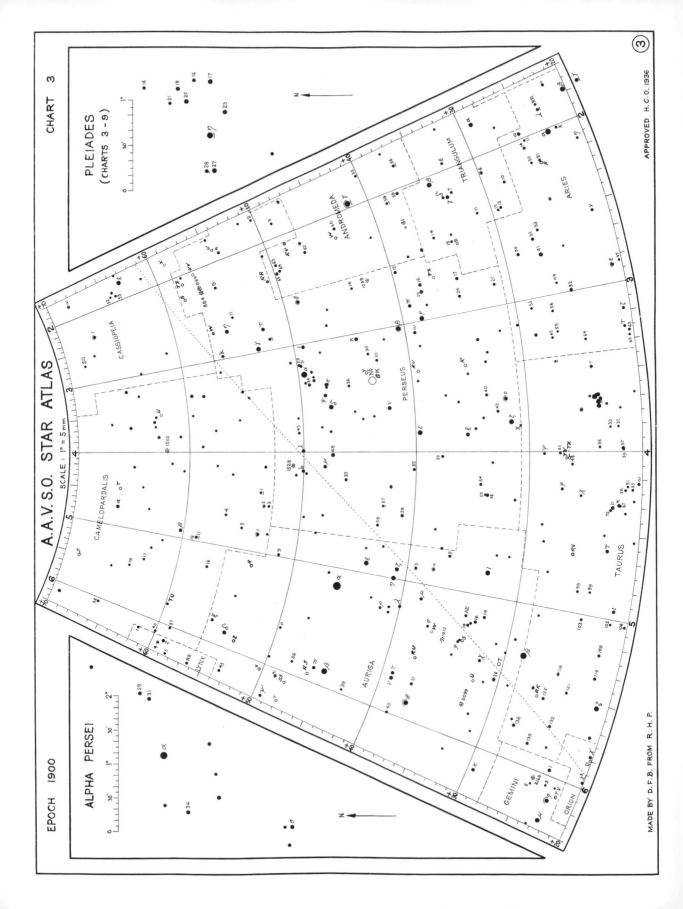

A.A.V.S.O. STAR ATLAS

CHART 3

EPOCH 1900

SCALE: 1° = 5 mm

PLEIADES (CHARTS 3 – 9)

ALPHA PERSEI

CASSIOPEIA

CAMELOPARDALIS

PERSEUS

ANDROMEDA

TRIANGULUM

ARIES

TAURUS

AURIGA

LYNX

GEMINI

ORION

APPROVED H.C.O. 1936

MADE BY D.F.B. FROM R.H.P.

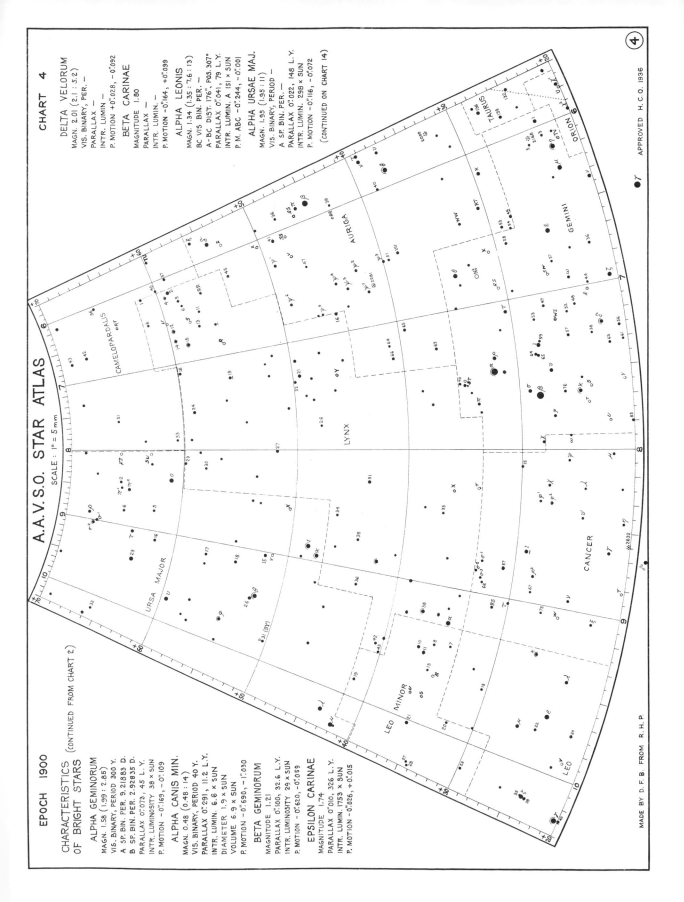

A.A.V.S.O. STAR ATLAS

SCALE : 1° = 5 mm

EPOCH 1900

CHART 4

APPROVED H.C.O. 1936

MADE BY D.F.B. FROM R.H.P.

CHARACTERISTICS OF BRIGHT STARS (CONTINUED FROM CHART 2)

ALPHA GEMINORUM
MAGN. 1.58 (1.99 : 2.85)
VIS. BINARY, PERIOD 300 Y.
A SP. BIN. PER. 9.2I883 D.
B SP. BIN. PER. 2.92835 D.
PARALLAX 0".073, 45 L.Y.
INTR. LUMINOSITY 38 × SUN
P. MOTION −0".169, −0".109

ALPHA CANIS MIN.
MAGN. 0.48 (0.48 : 14)
VIS. BINARY, PERIOD 40 Y.
PARALLAX 0".291, 11.2 L.Y.
INTR. LUMIN. 6.6 × SUN
DIAMETER 1.9 × SUN
VOLUME 6.9 × SUN
P. MOTION −0".690, −1".030

BETA GEMINORUM
MAGNITUDE 1.21
PARALLAX 0".100, 32.6 L.Y.
INTR. LUMINOSITY 29 × SUN
P. MOTION −0".620, −0".059

EPSILON CARINAE
MAGNITUDE 1.74
PARALLAX 0".010, 326 L.Y.
INTR. LUMIN. 1753 × SUN
P. MOTION −0".026, +0".015

DELTA VELORUM
MAGN. 2.01 (2.1 : 5.2)
VIS. BINARY, PER. —
PARALLAX —
INTR. LUMIN. —
P. MOTION +0".028, −0".092

BETA CARINAE
MAGNITUDE 1.80
PARALLAX —
INTR. LUMIN. —
P. MOTION −0".164, +0".099

ALPHA LEONIS
MAGN. 1.34 (1.35 : 7.6 : 13)
BC VIS BIN. PER. —
A−BC DIST. 176", P0S. 307°
PARALLAX 0".041, 79 L.Y.
INTR. LUMIN. A 151 × SUN
P. M. ABC −0".244, −0".001

ALPHA URSAE MAJ.
MAGN. 1.95 (1.95 : 11)
VIS. BINARY, PERIOD —
A SP. BIN. PER. —
PARALLAX 0".022, 148 L.Y.
INTR. LUMIN. 298 × SUN
P. MOTION −0".116, −0".072
(CONTINUED ON CHART 14)

CAMELOPARDALIS

URSA MAJOR

LYNX

AURIGA

GEMINI

CANCER

LEO MINOR

LEO

ORION

TAURUS

④

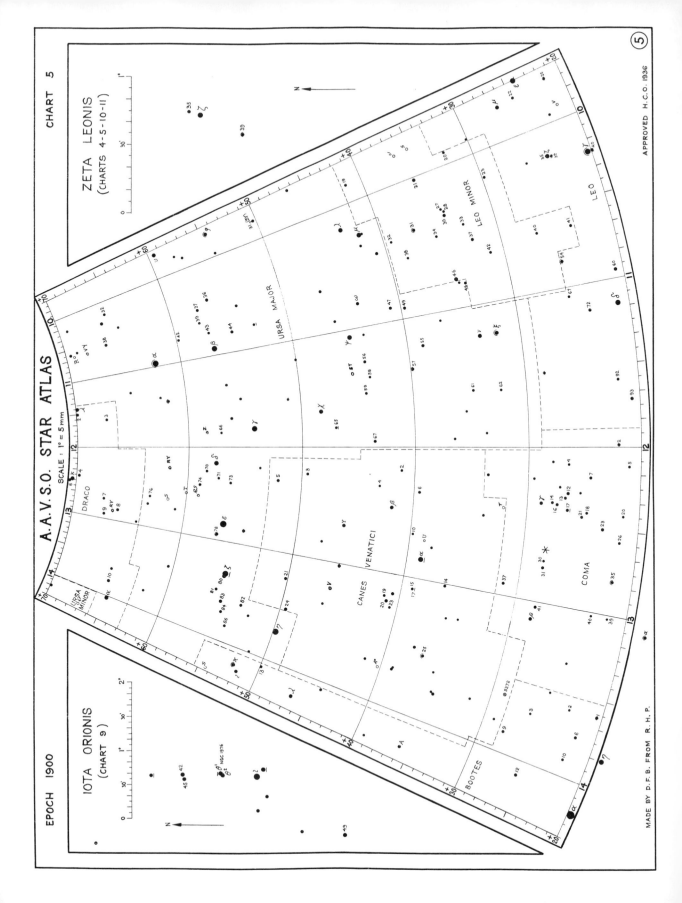

A.A.V.S.O. STAR ATLAS

EPOCH 1900

SCALE : 1° = 5 mm

CHART 5

ZETA LEONIS
(CHARTS 4-5-10-11)

IOTA ORIONIS
(CHART 9)

N

APPROVED H.C.O. 1936

MADE BY D.F.B. FROM R.H.P.

⑤

DRACO

URSA MAJOR

URSA MINOR

CANES VENATICI

COMA

LEO

LEO MINOR

BOOTES

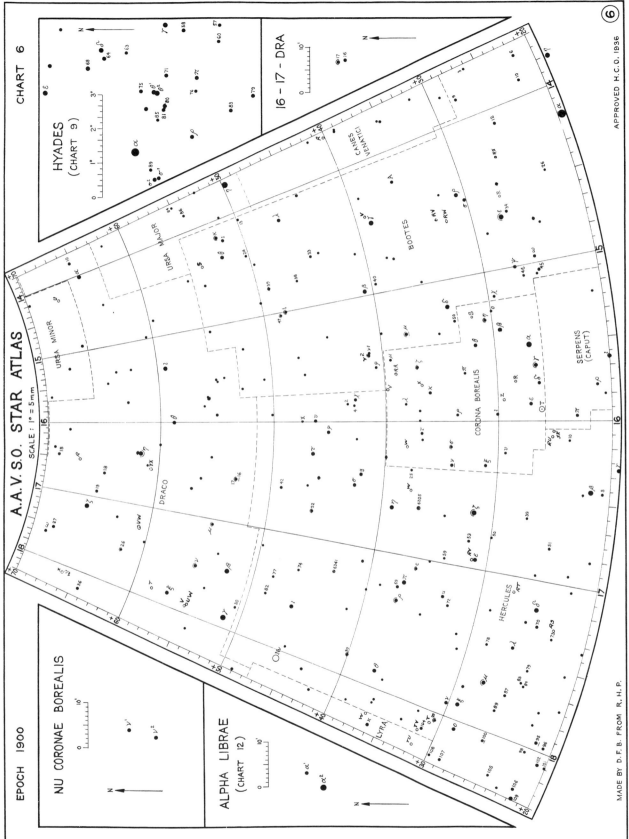

CHART 6

A.A.V.S.O. STAR ATLAS

EPOCH 1900

SCALE : 1° = 5 mm

HYADES
(CHART 9)

16 - 17 - DRA

NU CORONAE BOREALIS

ALPHA LIBRAE
(CHART 12)

URSA MINOR

URSA MAJOR

CANES
VENATICI

BOOTES

DRACO

CORONA BOREALIS

SERPENS
(CAPUT)

HERCULES

LYRA

APPROVED H.C.O. 1936

MADE BY D.F.B. FROM R.H.P.

⑥

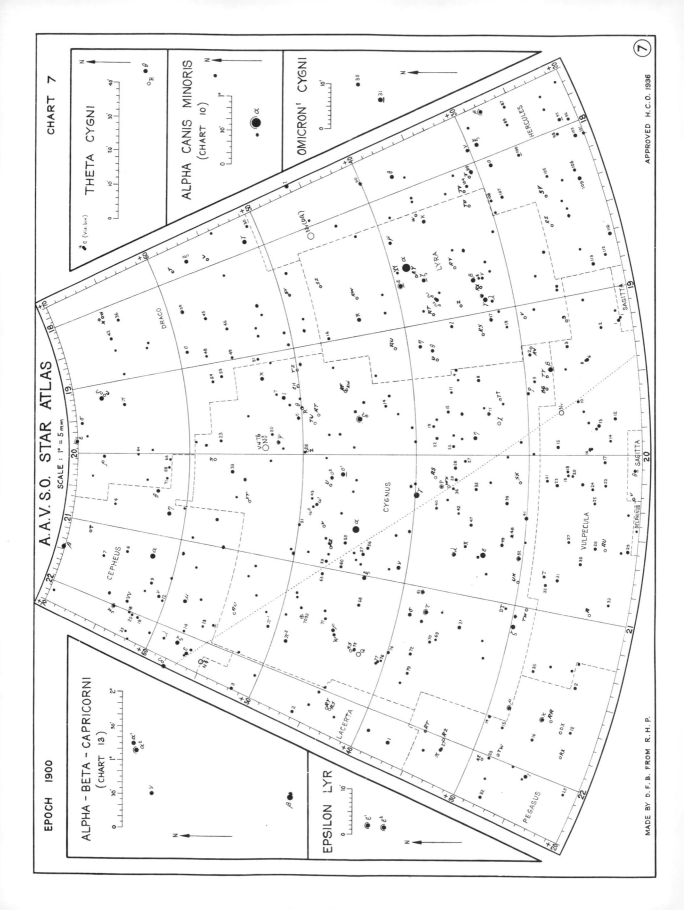

A.A.V.S.O. STAR ATLAS

SCALE : 1° = 5 mm

EPOCH 1900

CHART 7

THETA CYGNI

ALPHA CANIS MINORIS
(CHART 10)

OMICRON¹ CYGNI

ALPHA - BETA - CAPRICORNI
(CHART 13)

EPSILON LYR

APPROVED H.C.O. 1936

MADE BY D.F.B. FROM R.H.P.

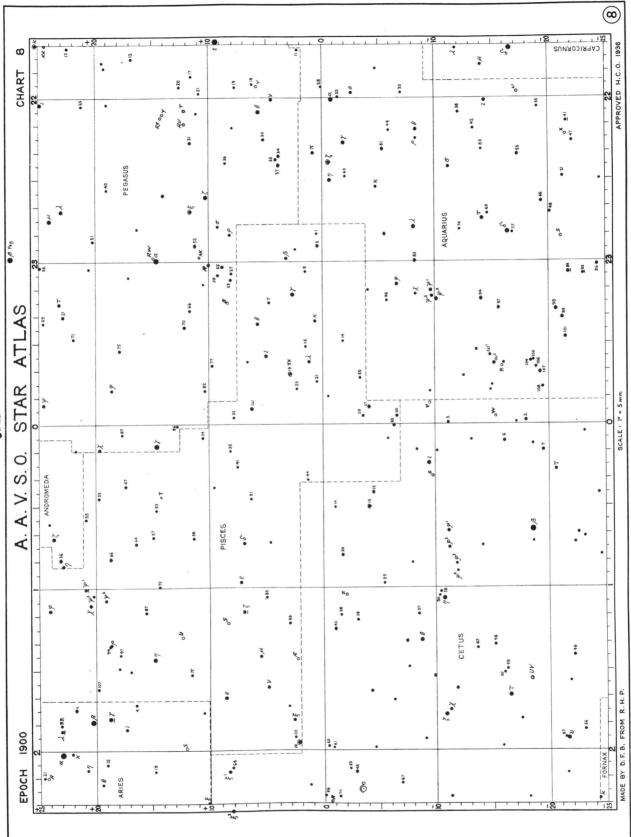

A.A.V.S.O. STAR ATLAS

EPOCH 1900

CHART 8

⑧

SCALE: 1° = 5 mm

APPROVED H.C.O. 1936

MADE BY D.F.B. FROM R.H.P.

PEGASUS

ANDROMEDA

PISCES

ARIES

AQUARIUS

CETUS

CAPRICORNUS

FORNAX

• β Peg

• α And

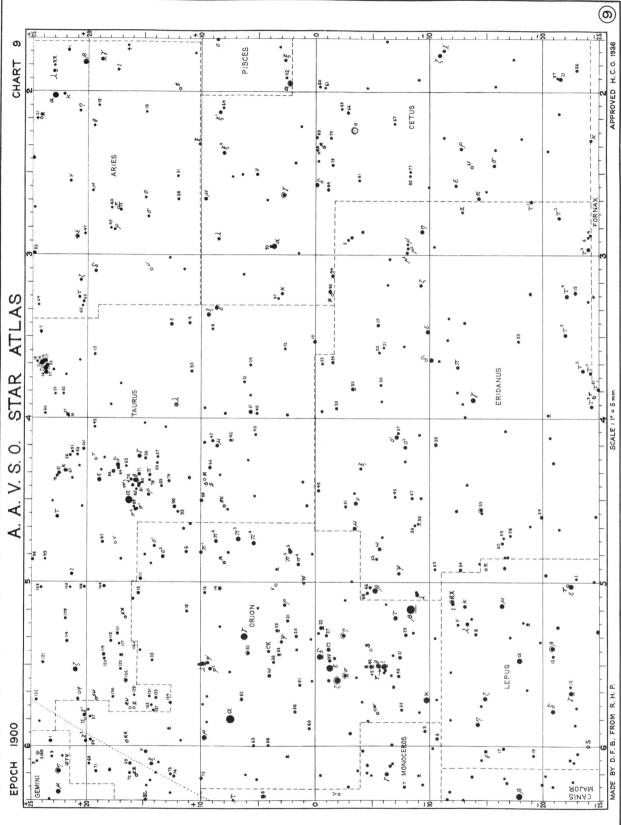

CHART 9

A.A.V.S.O. STAR ATLAS

EPOCH 1900

APPROVED H.C.O. 1936

SCALE : 1° = 5 mm

MADE BY D.F.B. FROM R.H.P.

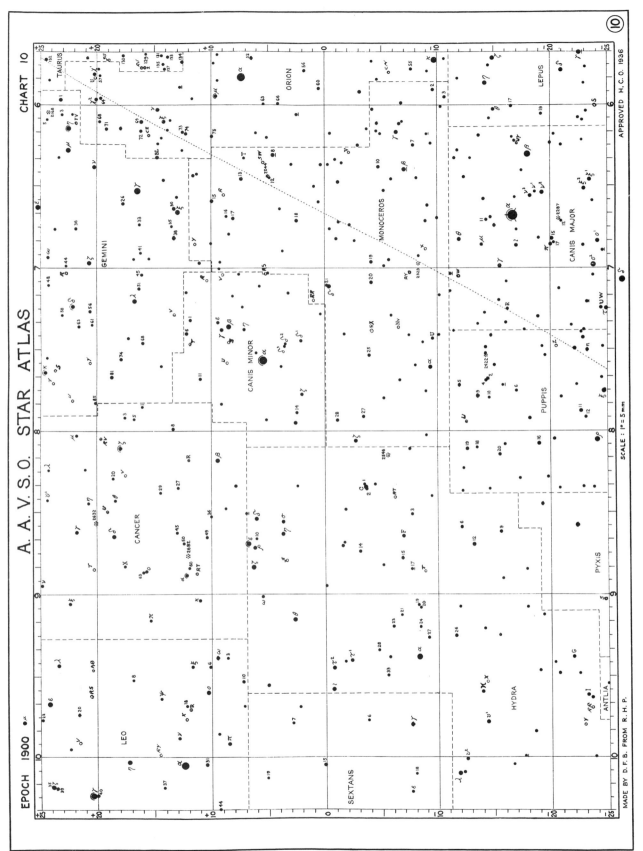

A.A.V.S.O. STAR ATLAS

EPOCH 1900

CHART 10

APPROVED H.C.O. 1936

SCALE : 1° = 5 mm

MADE BY D.F.B. FROM R.H.P.

TAURUS

ORION

GEMINI

MONOCEROS

CANIS MAJOR

LEPUS

CANIS MINOR

PUPPIS

CANCER

HYDRA

PYXIS

LEO

SEXTANS

ANTLIA

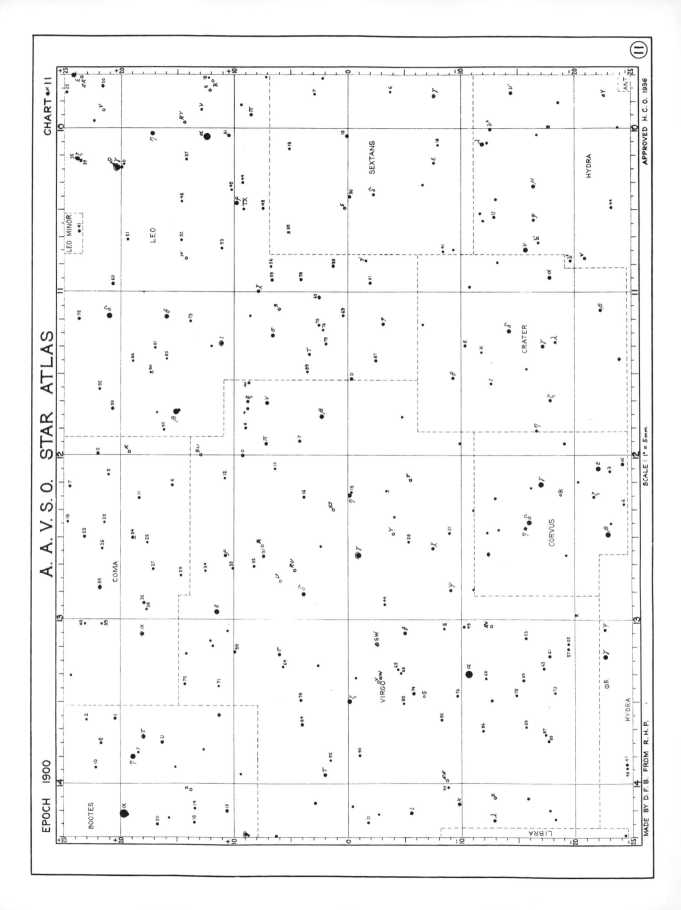

A. A. V. S. O. STAR ATLAS

EPOCH 1900

CHART №11

LEO MINOR

LEO

SEXTANS

HYDRA

ANT

COMA

CRATER

CORVUS

VIRGO

BOOTES

LIBRA

HYDRA

SCALE : 1° = 5mm

APPROVED H. C. O. 1936

MADE BY D. F. B. FROM R. H. P.

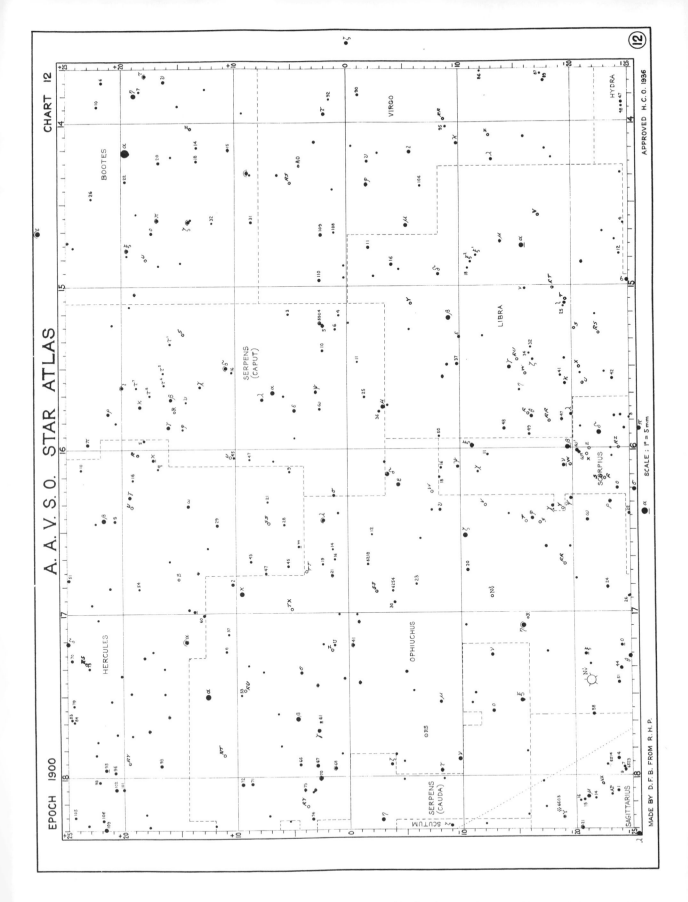

A.A.V.S.O. STAR ATLAS

EPOCH 1900

CHART 12

⑫

APPROVED H.C.O. 1936

SCALE : 1° = 5 mm

MADE BY D.F.B. FROM R.H.P.

BOOTES

HERCULES

SERPENS (CAPUT)

VIRGO

LIBRA

SCORPIUS

OPHIUCHUS

SERPENS (CAUDA)

SCUTUM

SAGITTARIUS

HYDRA

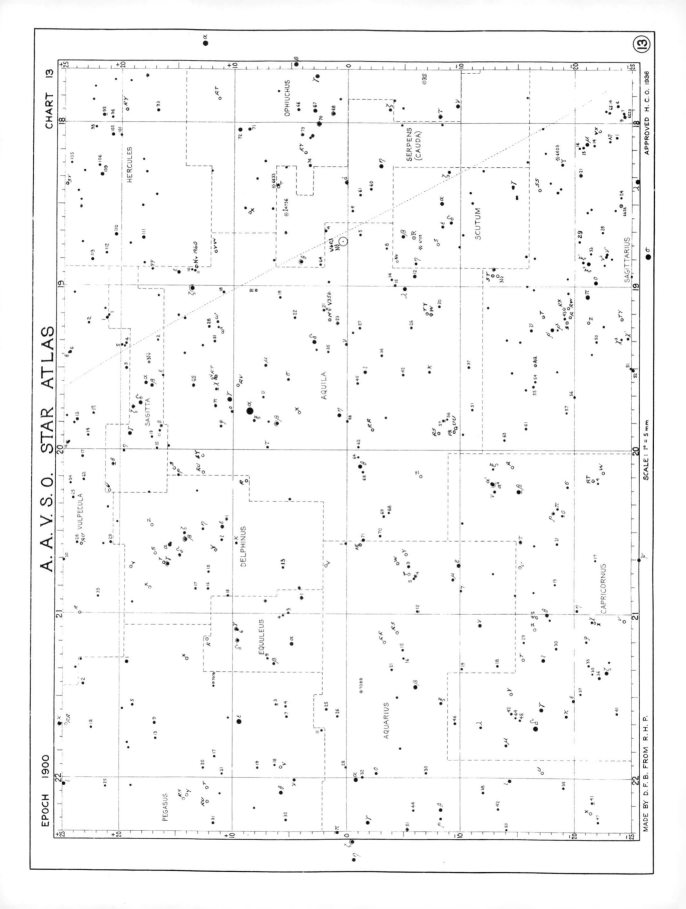

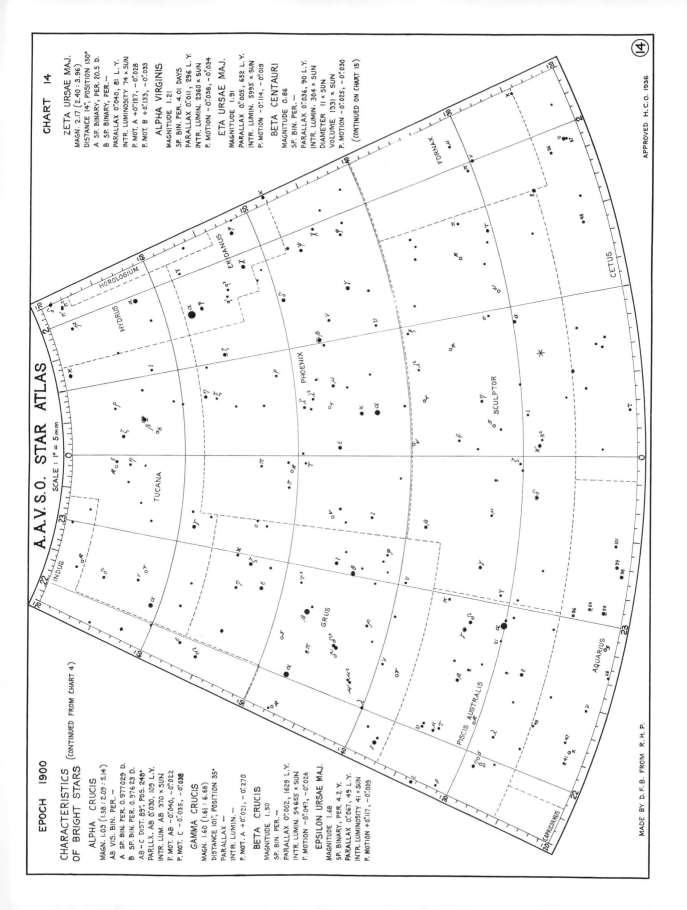

A.A.V.S.O. STAR ATLAS

SCALE : 1° = 5mm

EPOCH 1900

CHARACTERISTICS OF BRIGHT STARS (CONTINUED FROM CHART 4)

ALPHA CRUCIS
MAGN. 1.03 (1.58 : 2.09 : 5.14)
AB VIS. BIN. PER. —
A SP. BIN. PER. 0.977029 D.
B SP. BIN. PER. 0.976 23 D.
AB—C DIST. 89", POS. 248°
PARLLX. AB 0".030, 109 L.Y.
INTR. LUM. AB 370 × SUN
P. MOT. AB −0".040, −0".022
P. MOT. C −0".035, −0".038

GAMMA CRUCIS
MAGN. 1.60 (1.61 : 6.68)
DISTANCE 101", POSITION 35°
PARALLAX —
INTR. LUMIN. —
P. MOT. A +0".021, −0".270

BETA CRUCIS
MAGNITUDE 1.50
SP. BIN. PER. —
PARALLAX 0".002, 1629 L.Y.
INTR. LUMIN. 54 655 × SUN
P. MOTION −0".047, −0".026

EPSILON URSAE MAJ.
MAGNITUDE 1.68
SP. BINARY, PER. 4.2 Y.
PARALLAX 0".067, 49 L.Y.
INTR. LUMINOSITY 41 × SUN
P. MOTION +0".117, −0".009

CHART 14

ZETA URSAE MAJ.
MAGN. 2.17 (2.40 : 3.96)
DISTANCE 14", POSITION 150°
A SP. BINARY, PER. 20.5 D.
B SP. BINARY, PER. —
PARALLAX 0".040, 81 L. Y.
INTR. LUMINOSITY 74 × SUN
P. MOT. A +0".127, −0".028
P. MOT. B +0".133, −0".033

ALPHA VIRGINIS
MAGNITUDE 1.21
SP. BIN. PER. 4.01 DAYS
PARALLAX 0".011, 296 L. Y.
INTR. LUMIN. 2360 × SUN
P. MOTION −0".038, −0".034

ETA URSAE MAJ.
MAGNITUDE 1.91
PARALLAX 0".005, 652 L.Y.
INTR. LUMIN. 5995 × SUN
P. MOTION −0".114, −0".019

BETA CENTAURI
MAGNITUDE 0.86
SP. BIN. PER. —
PARALLAX 0".036, 90 L.Y.
INTR. LUMIN. 304 × SUN
DIAMETER 11 × SUN
VOLUME 1331 × SUN
P. MOTION −0".025, −0".030
(CONTINUED ON CHART 15)

APPROVED H.C.O. 1936

MADE BY D.F.B. FROM R.H.P.

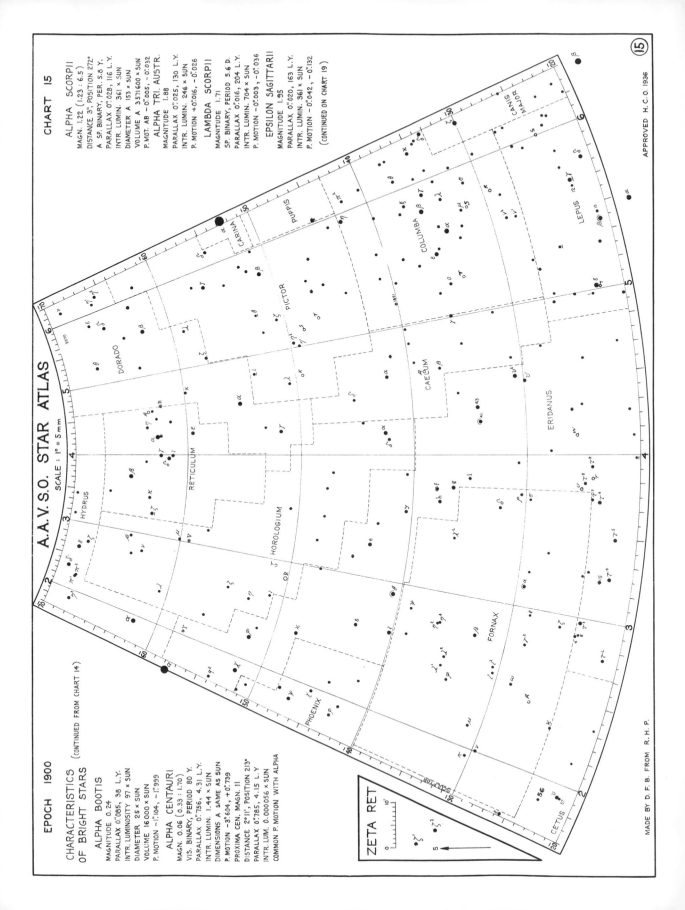

A.A.V.S.O. STAR ATLAS

SCALE : 1° = 5 mm

CHART 15

EPOCH 1900

CHARACTERISTICS (CONTINUED FROM CHART 14) OF BRIGHT STARS

ALPHA BOOTIS
MAGNITUDE 0.24
PARALLAX 0".085, 38 L.Y.
INTR. LUMINOSITY 97 × SUN
DIAMETER 25 × SUN
VOLUME 16 000 × SUN
P. MOTION −1".04, −1".999

ALPHA CENTAURI
MAGN. 0.06 (0.33 : 1.70)
VIS. BINARY, PERIOD 80 Y.
PARALLAX 0".756, 4.31 L.Y.
INTR. LUMIN. 1.44 × SUN
DIMENSIONS A SAME AS SUN
P. MOTION −3".604, +0".739
PROXIMA CEN. MAGN. 11
DISTANCE 2°11', POSITION 213°
PARALLAX 0".785, 4.15 L.Y
INTR. LUM. 0.000056 × SUN
COMMON P. MOTION WITH ALPHA

ALPHA SCORPII
MAGN. 1.22 (1.23 : 6.5)
DISTANCE 3", POSITION 272°
A SP. BINARY, PER. 5.8 Y.
PARALLAX 0".028, 116 L.Y.
INTR. LUMIN. 361 × SUN
DIAMETER 153 × SUN
VOLUME A 3 571 600 × SUN
P. MOT. AB −0".005, −0".032

ALPHA TRI. AUSTR.
MAGNITUDE 1.88
PARALLAX 0".025, 130 L.Y.
INTR. LUMIN. 246 × SUN
P. MOTION +0".016, −0".026

LAMBDA SCORPII
MAGNITUDE 1.71
SP. BINARY, PERIOD 5.6 D.
PARALLAX 0".016, 204 L.Y.
INTR. LUMIN. 704 × SUN
P. MOTION −0".003, −0".036

EPSILON SAGITTARII
MAGNITUDE 1.95
PARALLAX 0".020, 163 L.Y.
INTR. LUMIN. 361 × SUN
P. MOTION −0".042, −0".132

(CONTINUED ON CHART 19)

ZETA RET

APPROVED H.C.O. 1936

MADE BY D. F. B. FROM R. H. P.

15

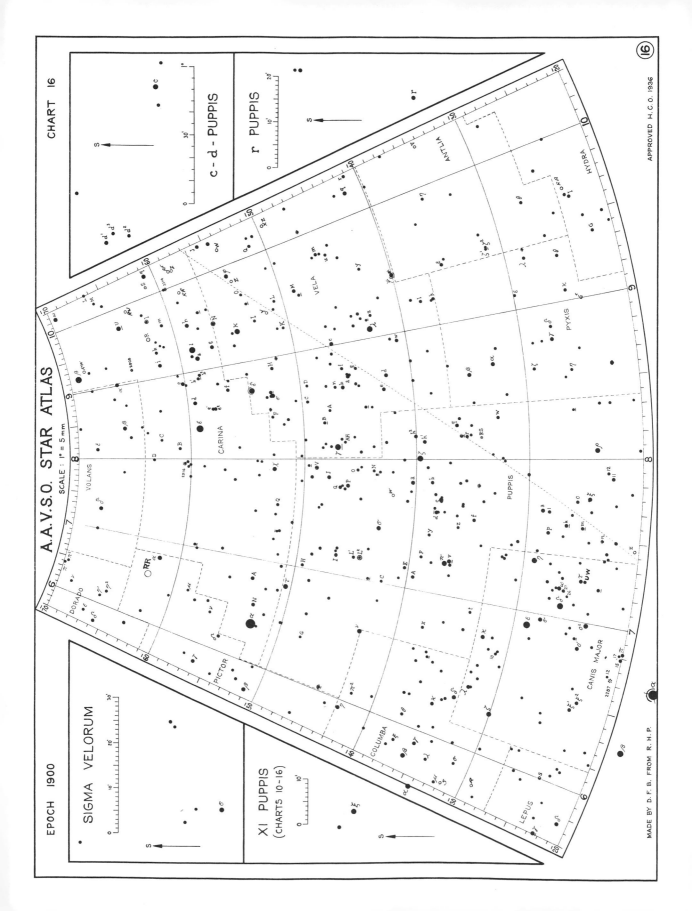

A.A.V.S.O. STAR ATLAS

EPOCH 1900

SCALE : 1° = 5 mm

CHART 16

CHART 16

c - d - PUPPIS

r PUPPIS

SIGMA VELORUM

XI PUPPIS
(CHARTS 10-16)

APPROVED H.C.O. 1936

MADE BY D.F.B. FROM R.H.P.

VOLANS

CARINA

VELA

ANTLIA

HYDRA

PYXIS

PUPPIS

DORADO

PICTOR

COLUMBA

LEPUS

CANIS MAJOR

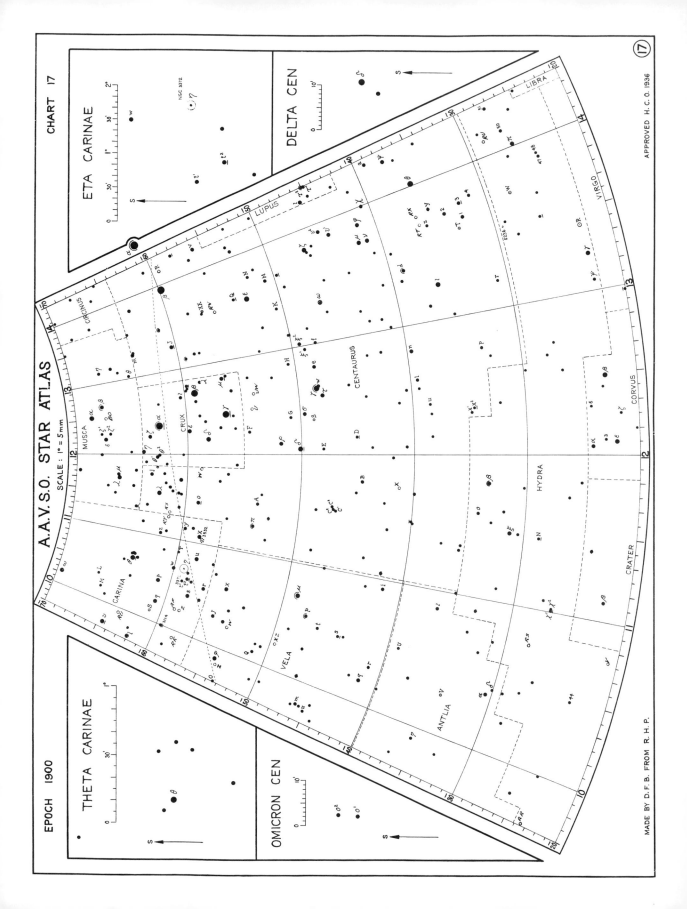

EPOCH 1900

A.A.V.S.O. STAR ATLAS

CHART 17

SCALE : 1° = 5mm

ETA CARINAE

DELTA CEN

THETA CARINAE

OMICRON CEN

APPROVED H. C. O. 1936

MADE BY D. F. B. FROM R. H. P.

CIRCINUS

LUPUS

MUSCA

CRUX

CENTAURUS

CARINA

VELA

ANTLIA

HYDRA

CRATER

CORVUS

VIRGO

LIBRA

NGC 3372

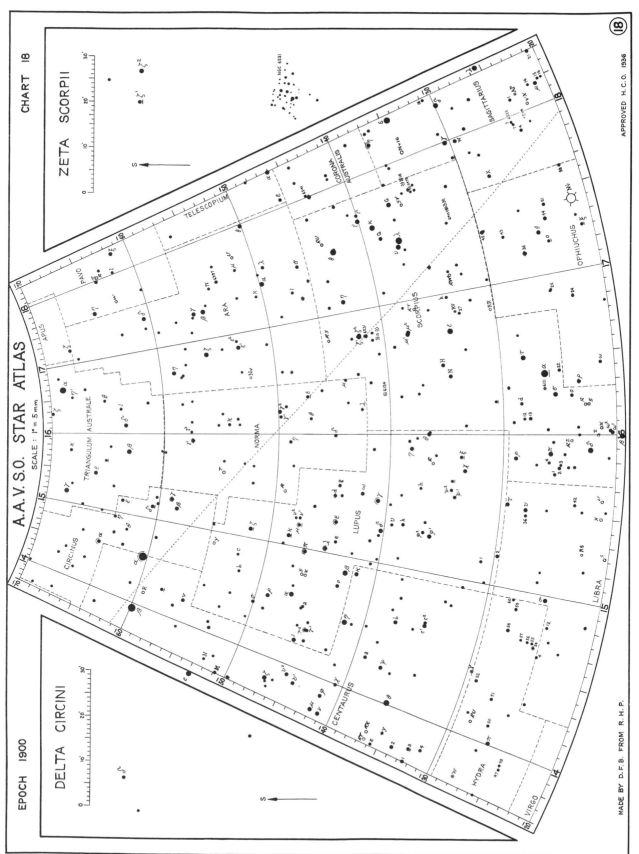

EPOCH 1900

A.A.V.S.O. STAR ATLAS

SCALE : 1° = 5 mm

CHART 18

ZETA SCORPII

DELTA CIRCINI

APPROVED H.C.O. 1936

MADE BY D.F.B. FROM R.H.P.

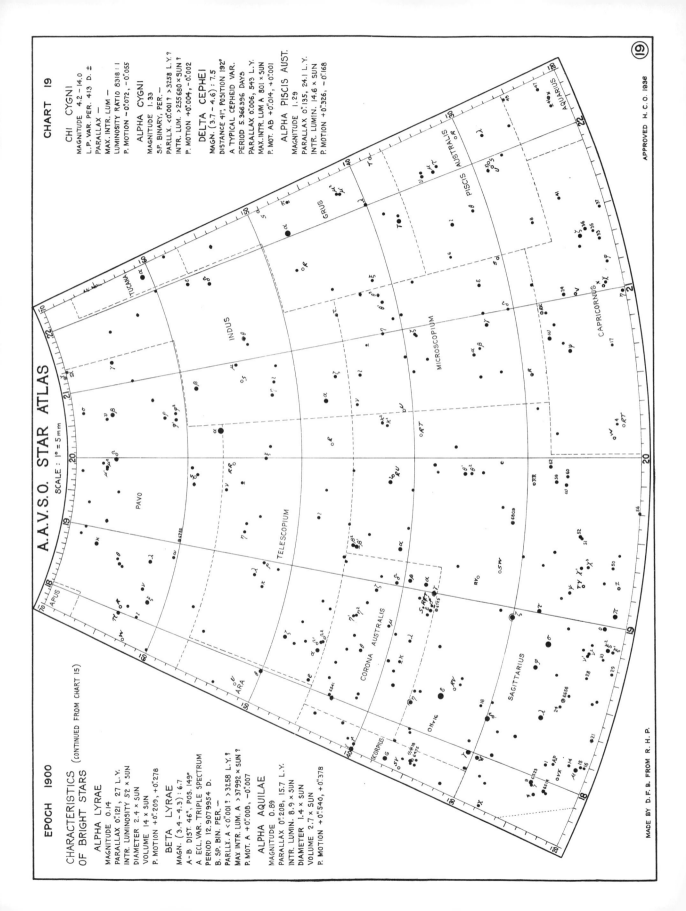

A.A.V.S.O. STAR ATLAS

EPOCH 1900

SCALE : 1° = 5 mm

CHART 19

CHARACTERISTICS OF BRIGHT STARS (CONTINUED FROM CHART 15)

ALPHA LYRAE
MAGNITUDE 0.14
PARALLAX 0".121, 27 L.Y.
INTR. LUMINOSITY 52 × SUN
DIAMETER 2.4 × SUN
VOLUME 14 × SUN
P. MOTION +0".209, +0".278

BETA LYRAE
MAGN. (3.4 − 4.3) : 6.7
A-B DIST. 46", POS. 149°
A ECL. VAR. TRIPLE SPECTRUM
PERIOD 12.90 9954 D.
B. SP. BIN. PER. —
PARLLX. A < 0".001 ? > 3258 L.Y. ?
MAX INTR. LUM. A > 37992 × SUN ?
P. MOT. A +0".008, −0".007

ALPHA AQUILAE
MAGNITUDE 0.89
PARALLAX 0".208, 15.7 L.Y.
INTR. LUMIN. 8.9 × SUN
DIAMETER 1.4 × SUN
VOLUME 2.7 × SUN
P. MOTION +0".540, +0".378

CHI CYGNI
MAGNITUDE 4.2 − 14.0
L.P. VAR. PER. 413 D. ±
PARALLAX —
MAX. INTR. LUM. —
LUMINOSITY RATIO 8318 : I
P. MOTION −0".072, −0".055

ALPHA CYGNI
MAGNITUDE 1.33
SP. BINARY, PER. —
PARLLX. < 0".001 ? > 3258 L.Y. ?
INTR. LUM. > 255680 × SUN ?
P. MOTION +0".004, −0".002

DELTA CEPHEI
MAGN. (3.7 − 4.6) : 7.5
DISTANCE 41", POSITION 192°
A TYPICAL CEPHEID VAR.
PERIOD 5.366396 DAYS
PARALLAX 0".006, 543 L.Y.
MAX. INTR. LUM. A 801 × SUN
P. MOT. AB +0".014, +0".001

ALPHA PISCIS AUST.
MAGNITUDE 1.29
PARALLAX 0".135, 24.1 L.Y.
INTR. LUMIN. 14.6 × SUN
P. MOTION +0".326, −0".168

MADE BY D.F.B. FROM R.H.P.

APPROVED H.C.O. 1938

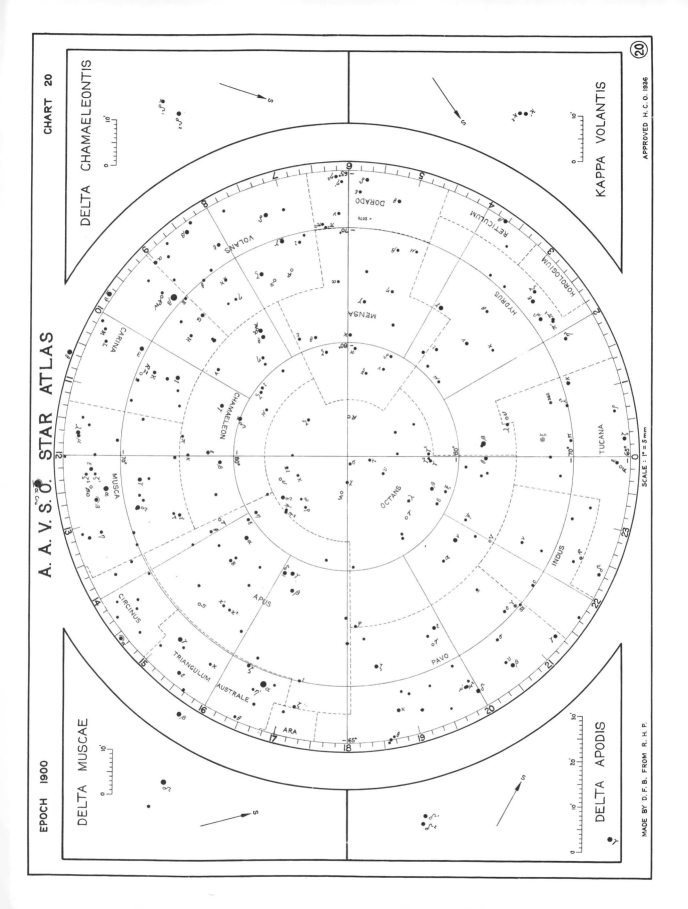

A.A.V.S.O. STAR ATLAS

EPOCH 1900

CHART 20

DELTA CHAMAELEONTIS

KAPPA VOLANTIS

DELTA MUSCAE

DELTA APODIS

SCALE : 1" = 5 mm

APPROVED H.C.O. 1936

MADE BY D.F.B. FROM R.H.P.

Acknowledgments

The following illustrations were reproduced from *Introduction to Astronomy*, 2nd ed., by Cecilia Payne-Gaposchkin and Katherine Haramundanis (Englewood Cliffs: Prentice-Hall; © 1970, 1954 by Prentice-Hall, Inc., Englewood Cliffs, New Jersey): 1.4, 1.5, 1.6, 2.3, 2.5, 4.14, 5.2, 8.1, 8.13, 8.14, 12.1, 13.3, 14.9, 16.1, 16.4, 16.8, 16.12, 16.13. Illustrations from the *Astrophysical Journal*, published by the University of Chicago Press, are copyrighted © 1946, 1946, 1954, 1956, 1958, 1962, 1964, 1973, 1974, 1976 by the American Astronomical Society. Figures 3.2 and 5.4 are reproduced by permission of the University of Chicago Press (© 1963 and © 1943 by the University of Chicago Press). Lola Chaisson, Beryl Langer, and June Wallace prepared color–magnitude diagrams and other drawings for the volume.

General Index

Index of Stars

Index of Clusters and Galaxies

h Persei: 24, 47, 60, 61, 87, 90, 98,
  105
Perseus association: 60, 61, 87, 90,
  98, 128
Pleiades: 13, 16, 24, 46, 47, 60, 80,
  83, 84, 125, 147
Praesepe: 13, 125, 130, 132, 143,
  200, 203
I Puppis: 113

Sculptor system: 15
Small Magellanic cloud: 123

Taurus association: 55, 56, 58, 59,
  109
Trumpler
  24: 149
  25: 149

26: 149
29: 149
30: 149
47 Tucanae: 162, 167, 168, 169, 171,
  175, 179

Ursa Major cluster: 13, 76, 81, 83,
  91